Irvine, John Maxwell.
Heavy nuclei, superheavy nuclei, and neutron stars / J. M. Irvine. — Oxford [Eng.] : Clarendon Press, 1975.
164 p. : ill. ; 24 cm. — (Oxford studies in nuclear physics) GB***
Bibliography: p. [159]-160.
Includes indexes.
ISBN 0-19-851510-3 : £8.00

1. Nuclear shell theory. 2. Neutron stars. I. Title. II. Series.
QC793.3.S8I78 539.7'43 75-327634
MARC

OXFORD STUDIES
IN
NUCLEAR PHYSICS

GENERAL EDITOR
P. E. HODGSON

HEAVY NUCLEI, SUPERHEAVY NUCLEI, AND NEUTRON STARS

J. M. Irvine

CLARENDON PRESS · OXFORD
1975

Oxford University Press, Ely House, London W. 1

GLASGOW NEW YORK TORONTO MELBOURNE WELLINGTON
CAPE TOWN IBADAN NAIROBI DAR ES SALAAM LUSAKA ADDIS ABABA
DELHI BOMBAY CALCUTTA MADRAS KARACHI LAHORE DACCA
KUALA LUMPUR SINGAPORE HONG KONG TOKYO

ISBN 0 19 851510 3

Text set in 10/12 pt. IBM Press Roman, printed by photolithography,
and bound in Great Britain at The Pitman Press, Bath

PREFACE

It could be argued that the trigger which set off the explosion of interest in the physical sciences since the second World War was the possibility of harnessing the enormous energy released in nuclear fission. Yet despite the pioneering work of Bohr and Wheeler a long time elapsed before it became possible to explain the detailed fission mechanism, and only in recent years have many of the outstanding questions in this field been clarified. At the same time the whole question of nuclear stability and the possible existence of relatively long-lived superheavy elements has been the subject of deep scrutiny. Many of the experiments suggested to test the new theories involve heavy ion reactions at energies close to, or above, the Coulomb barrier and the consequent need to design and build a new generation of heavy ion accelerators with which to carry out these experiments.

Paralleling these developments in terrestrial nuclear physics the discovery of pulsars and the suggestion that they may well be neutron stars has once more emphasized the importance of understanding nuclear processes in astrophysics. It thus appears a timely exercise to prepare a brief, unified introduction to these matters. Many of the ideas related to the topics discussed are in state of flux. Thus in no way is this presentation intended to be definitive. I have enclosed a few references, and by and large these are not original research papers but rather reviews which I have found particularly readable and through which the reader will find a more extensive bibliography of original material.

I would like to acknowledge many helpful discussions with my colleagues in Manchester and in particular Dr. C. Pwu. Most of the results presented in Chapter 7 are taken from his Ph.D. thesis and the results in Chapters 8 and 9 relied heavily on computer codes for which he was responsible. I am grateful to Miss V. Harney for the most efficient way she translated my often unintelligible scrawl into a legible typescript.

Finally, I would like to dedicate this book to Ritchie Middlemass, a patient and considerate teacher.

Manchester 1974. J. M. I.

CONTENTS

1.	INTRODUCTION	1
2.	THE OBSERVED SYSTEMATICS OF HEAVY NUCLEI	3
3.	NUCLEAR MODELS	23
	3.1. Introduction	23
	3.2. The liquid-drop model	23
	3.3. The individual-particle model	40
4.	NUCLEAR CORRELATIONS	78
	4.1. Introduction	78
	4.2. Hastree–Fock theory: static and time-dependent	78
	4.3. Idealized residual interactions	89
5.	COLLECTIVE ROTATIONS	103
6.	NUCLEAR STABILITY AND SHELL EFFECTS	113
7.	FISSION AND ALPHA DECAY	123
8.	SUPERHEAVY NUCLEI	142
9.	NEUTRON STARS	150
	REFERENCES	159
	AUTHOR INDEX	161
	SUBJECT INDEX	163

1

INTRODUCTION

Our topic is the structure of heavy, superheavy, and ultra-superheavy (i.e. neutron-star) nuclei. That is, we are interested in *nuclear matter.* Not in the rather academic sense of an infinite equal number of neutrons and protons in which the Coulomb force has mysteriously been switched off, which technically is the meaning of the term 'nuclear matter', but rather in the sense of the possible states of matter composed of nucleons at densities comparable to, or greater than, those found in terrestial atomic nuclei (Bethe 1971; Irvine 1972).

Where we draw the line between what is heavy and what is not, is, of course, rather arbitrary. We shall draw the line at ^{208}Pb, and our study will concentrate on nuclei with $A \geqslant 208$ and $Z \geqslant 82$. We shall be particularly interested in those phenomena which are essentially statistical in origin, reflecting the participation of very many particles, e.g. collective rotations, fission, shape isomers, superfluidity, etc. This is not to say that some or all of these effects are not seen in lighter nuclei. In particular, the discussion of collective rotations would be applicable to an account of the rare-earth nuclei. Indeed we shall briefly discuss the phenomena of the 'back-bending' of moments of inertia (Sorensen 1973) which to date have only been observed in the rare-earth nuclei but which, if our current ideas have any validity, should also be displayed in the heavy actinides.

In our region of interest there are only ten naturally occurring elements,† although there are many more isotopes, and if this was all there was to discuss we would indeed be severely restricted. However, since the detonation of the first atomic bomb man has been manufacturing new nuclei, and there are currently fourteen man-made elements and their associated isotopes which can be added to our catalogue. One of the subjects we shall discuss concerns the stability of these very heavy nuclei and of the possible existence of an 'island' of stability of superheavy elements (Nix and Swiatecki 1965). This leads naturally to the suggestion that there might exist a continent of ultra-superheavy elements whose mass is so great that their self-gravitational interaction is the dominant stabilizing influence. In order to produce such enormous gravitational potentials these giant nuclei would require to be of stellar masses. Extrapolating from the growing neutron excess in heavy nuclei we conclude that these giant nuclei would be composed mostly of neutrons, and hence the term 'neutron stars' (Harrison, Thorne, Wakano, and Wheeler 1965). Although the existence of such strange objects was postulated by Landau in 1932 and has been strongly advocated by Zwicky since 1934, it was not until the discovery by Hewish in 1968 of 'pulsars' that a

†Most of which are short lived decay products of longer lived uranium isotopes.

possible identification of neutron stars with any observed astronomical object became likely.

Thus our subject matter covers a wide range of physical phenomena. Such short volume as this cannot attempt to be definitive. Instead we have attempted to outline some of the more outstanding properties of the known heavy nuclei and to develop the theories which have been used to describe them. We have then used those theories to discuss some of the possible properties of superheavy elements and neutron stars.

The possible production of superheavy elements has not been discussed. Great hopes have been pinned on producing superheavies in heavy-ion fusion reactions and these require high-energy heavy-ion accelerators which are only now under construction. In fairness it should be stated that at present the indications are that the fusion cross-sections are likely to be disappointingly small. It is just possible that superheavy elements may be synthesized in neutron stars and may subsequently be found in meteorites or cosmic rays.

Our discussion of neutron stars is extremely brief. Much current work is extremely speculative. However, it is probably true to say that there is a greater richness of physical phenomena possibly associated with neutron stars than with any other objects in nature. We have not discussed star quakes nor pulsing mechanisms.

What we have attempted to do is to develop certain central ideas and provide the reader with the necessary introduction which will allow him to follow the current literature. There are many variations on the theories we have developed and, with apologies to the authors of these variations, we have claimed that these are minor perturbations on our main theme. In order not to confuse the reader we have not developed all the variations simultaneously. The path we have followed involves a personal view of the subject, but as always this is the author's perogative.

2

THE OBSERVED SYSTEMATICS OF HEAVY NUCLEI

Fig. 2.1 shows a chart of the observed nuclides in the plane of the neutron number N and the proton number Z. In this Figure we indicate the naturally occurring isotopes and the principal decay modes of the short-lived isotopes. It is immediately obvious that the region of heavy nuclei is characterized by three novel features not typical of the remainder of the chart:

(1) alpha decay becomes the dominant decay mechanism for an increasing number of nuclei;

(2) amongst the heaviest observed nuclei spontaneous fission becomes important, and in a few cases it rivals alpha decay as the dominant decay mechanism;

(3) the density of naturally occurring nuclides falls off rapidly with increasing mass number.

In Fig. 2.2 is shown the region for $A > 208$ in considerably more detail, and we immediately notice that it is populated almost entirely by unstable nuclei. The only exceptions to this are ^{208}Pb which is completely stable and ^{209}Bi which, although we might expect it to be alpha-particle unstable (see p. 113), has an observed half-life in excess of 2×10^{18} years and hence, to all intents and purposes, may be considered to be completely stable. We can consider these two isotopes to be anomalies in our region of interest, and their principal value to us will be as sources of information which may provide us with a key to the understanding of their heavier neighbours.

It is well known that the measured masses of nuclei are not equal to the sum of the masses of the particles of which they are composed. The missing mass is interpreted as the binding energy of the nucleus $B(A,Z)$,

$$B(A, Z) = \Delta M(A, Z)c^2 = (Nm_n + Zm_p - M(A,Z))c^2 \qquad (2.1)$$

In Fig. 2.3 we plot the observed value of B_{max}/A as a function of mass number A, where B_{max} is the binding energy of the most stable nucleus of a given value of A. We see that for nuclei with masses greater than ^{56}Fe there is a general trend towards a declining binding energy per particle. This decline shows a marked acceleration beyond ^{208}Pb. In Table 2.1 we present some more detailed information on the binding energies, lifetimes, and decay modes of the heavy nuclei.

Fig. 2.4 is a contour diagram of the binding energy per particle of the nuclides in the N, Z-plane. We see that the diagram resembles a peninsula in which the naturally occurring isotopes form a mountain ridge with peaks corresponding to particularly stable nuclei, and the artificially produced radioactive isotopes

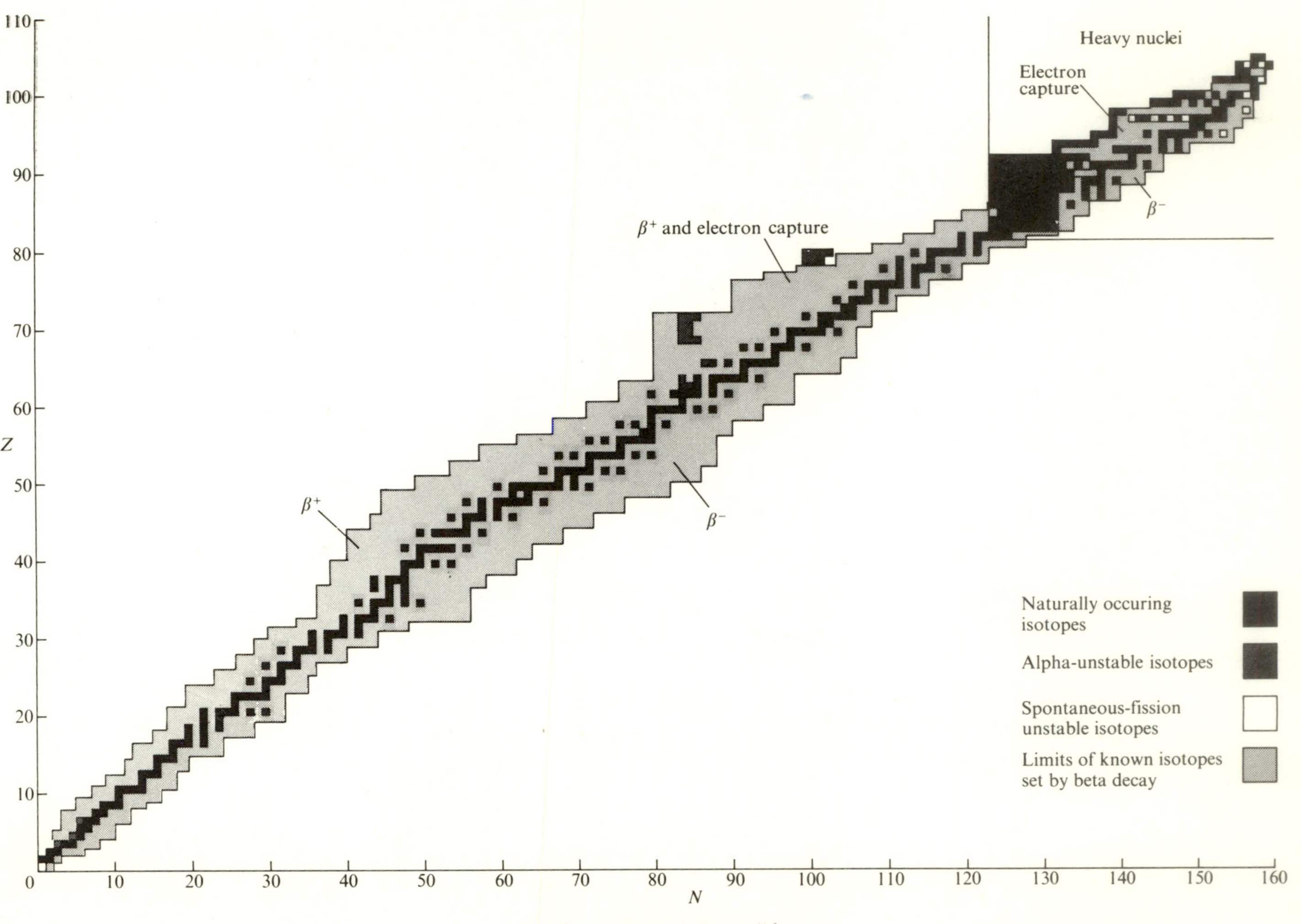

FIG. 2.1. Chart of the nuclides.

appear at progressively greater depths as their binding energies decrease. A central problem in modern nuclear physics concerns the possible existence of off-shore islands, i.e. either stable or near-stable exotic nuclei, with neutron excesses which are much greater or much less than those observed along the peninsula, or superheavy elements which appear as off-shore continuations of the peninsula. If we assume that pulsars are indeed neutron stars then we could claim to have observed a reappearance of the peninsula of stability at $A \gtrsim 10^{55}$. Rather a new continent, dominated by gravitational effects, than another island in the archipelego.

Throughout the periodic table it is observed than even–even nuclei are more stable than either their odd-A or odd–odd neighbours. In addition, the ground-state spins and parities are invariably 0^+. This relative stability is even more pronounced in the heavy nuclei (see Table 2.1). The observed low-lying spectra of the most stable even–even nuclei as a function of mass number A is plotted in Fig. 2.5. This Figure shows the truly remarkable result that for $A \gtrsim 230$ the spectra becomes independent of mass number. There is no other region of the periodic table where such an independence holds for such a wide range of mass numbers.

It is usual in a weakly interacting many-fermion system to think of adding particles to states at the top of the Fermi sea and to consider the low-lying excitation spectrum to be provided by the interaction between such particles near the Fermi surface. If this were the case the spectra would depend critically on the mass number, as indeed it does for $A \lesssim 220$ and for all odd-A nuclei. However, for a strongly interacting system it is possible to spontaneously generate collective excitations of the system in which a large fraction of the particles behave coherently. Such a spectrum would then be insensitive to small fluctuations in the mass number (note that in going from $A = 230$ to $A = 250$ the mass number has changed by less than 10 per cent). This view of the nature of the excitation spectrum of the heavy nuclei is reinforced by the observation that the spectrum closely resembles that of a quantal rigid rotator,

$$E_J = J(J+1)\hbar^2/2\mathscr{I}_{\text{eff}}, \tag{2.2}$$

where E_J is the excitation energy of the state of angular momentum J and $\mathscr{I}_{\text{eff}}$ is the effective moment of inertia. The constancy of the spectra for $A \gtrsim 230$ indicates that the effective moment of inertia is constant. With $E_2 = 0{\cdot}04$ MeV we deduce that $\mathscr{I}_{\text{eff}}$ has the value

$$\mathscr{I}_{\text{eff}} = 6\hbar^2/2E_2 = 3{\cdot}25 \times 10^{-41} \text{ MeV s}^2. \tag{2.3}$$

This may be compared with the classical rigid-body moment of inertia obtained by considering the nucleus to be a sphere of mass M and radius $R = A^{\frac{1}{3}} r_0$,

$$\mathscr{I}_{\text{rigid}} \simeq A^{\frac{5}{3}} \times 10^{-44} \text{ MeV s}^2. \tag{2.4}$$

For $A \approx 250$ we find that $\mathscr{I}_{\text{rigid}} \approx 3\ \mathscr{I}_{\text{eff}}$. In Chapter 5 we shall use this discrepancy between the rigid-body value of the moment of inertia and the effective

FIG. 2.2. Detailed chart of the heavy nuclides.

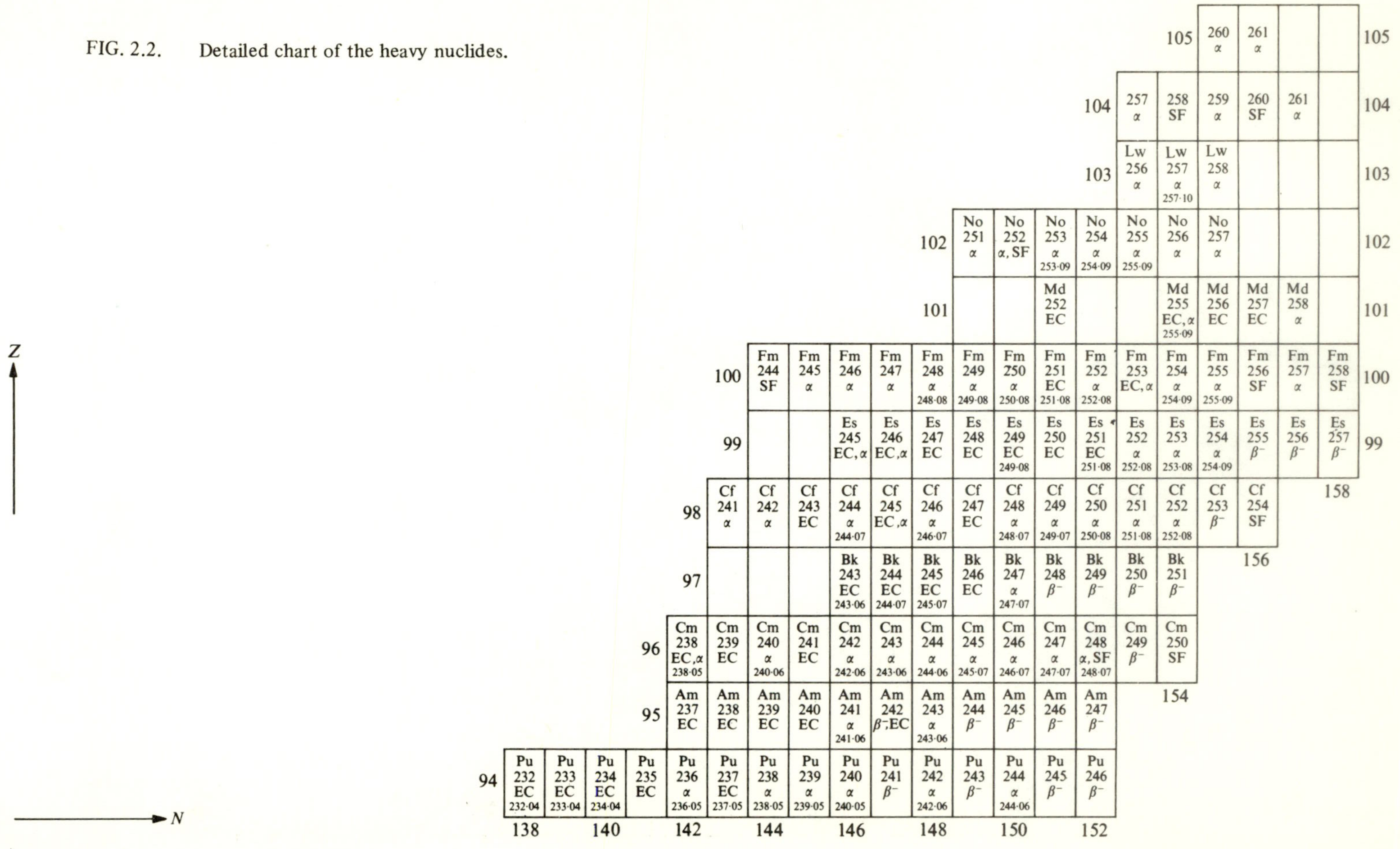

Z \ N	138		140		142		144		146		148		150		152		154		156		158
105																		260 α	261 α		
104																257 α	258 SF	259 α	260 SF	261 α	
103																Lw 256 α	Lw 257 α 257·10	Lw 258 α			
102												No 251 α	No 252 α, SF	No 253 α 253·09	No 254 α 254·09	No 255 α 255·09	No 256 α	No 257 α			
101														Md 252 EC			Md 255 EC, α 255·09	Md 256 EC	Md 257 EC	Md 258 α	
100							Fm 244 SF	Fm 245 α	Fm 246 α	Fm 247 α	Fm 248 α 248·08	Fm 249 α 249·08	Fm 250 α 250·08	Fm 251 EC 251·08	Fm 252 α 252·08	Fm 253 EC, α	Fm 254 α 254·09	Fm 255 α 255·09	Fm 256 SF	Fm 257 α	Fm 258 SF
99									Es 245 EC, α	Es 246 EC, α	Es 247 EC	Es 248 EC	Es 249 EC 249·08	Es 250 EC	Es 251 EC 251·08	Es 252 α 252·08	Es 253 α 253·08	Es 254 α 254·09	Es 255 β^-	Es 256 β^-	Es 257 β^-
98						Cf 241 α	Cf 242 α	Cf 243 EC	Cf 244 α 244·07	Cf 245 EC, α	Cf 246 α 246·07	Cf 247 EC	Cf 248 α 248·07	Cf 249 α 249·07	Cf 250 α 250·08	Cf 251 α 251·08	Cf 252 α 252·08	Cf 253 β^-	Cf 254 SF		
97									Bk 243 EC 243·06	Bk 244 EC 244·07	Bk 245 EC 245·07	Bk 246 EC	Bk 247 α 247·07	Bk 248 β^-	Bk 249 β^-	Bk 250 β^-	Bk 251 β^-				
96					Cm 238 EC, α 238·05	Cm 239 EC	Cm 240 α 240·06	Cm 241 EC	Cm 242 α 242·06	Cm 243 α 243·06	Cm 244 α 244·06	Cm 245 α 245·07	Cm 246 α 246·07	Cm 247 α 247·07	Cm 248 α, SF 248·07	Cm 249 β^-	Cm 250 SF				
95					Am 237 EC	Am 238 EC	Am 239 EC	Am 240 EC	Am 241 α 241·06	Am 242 β^-, EC	Am 243 α 243·06	Am 244 β^-	Am 245 β^-	Am 246 β^-	Am 247 β^-						
94	Pu 232 EC 232·04	Pu 233 EC 233·04	Pu 234 EC 234·04	Pu 235 EC	Pu 236 α 236·05	Pu 237 EC 237·05	Pu 238 α 238·05	Pu 239 α 239·05	Pu 240 α 240·05	Pu 241 β^-	Pu 242 α 242·06	Pu 243 β^-	Pu 244 α 244·06	Pu 245 β^-	Pu 246 β^-						

Z \ N	126		128		130		132		134		136		138		140		142		144		146		148
93										Np 228 SF?	Np 229 α	Np 230 α 230·04	Np 231 α 231·04	Np 232 EC	Np 233 EC 233·04	Np 234 EC	Np 235 EC	Np 236 EC,β^- 236·05	Np 237 α	Np 238 β^- 238·05	Np 239 β^-	Np 240 β^-	Np 241 β^-
92										U 227 α 227·03	U 228 α 228·03	U 229 EC,α 229·03	U 230 α 230·03	U 231 EC	U 232 α 232·04	U 233 α 233·04	U 234 α 234·04	U 235 α 235·04	U 236 α 236·05	U 237 β^-	U 238 α 238·05	U 239 β^-	U 240 β^-
91								Pa 224 α	Pa 225 α 225·03	Pa 226 α, EC 226·03	Pa 227 α, EC 227·03	Pa 228 EC 228·03	Pa 229 EC	Pa 230 EC,β^-	Pa 231 α 231·04	Pa 232 β^-	Pa 233 β^-	Pa 234 β^- 234·04	Pa 235 β^-	Pa 236 β^-	Pa 237 β^-		
90	Th 216 α	Th 217 α	Th 218 α	Th 219 α	Th 220 α	Th 221 α	Th 222 α	Th 223 α 223·02	Th 224 α 224·02	Th 225 α, EC 225·02	Th 226 α 226·02	Th 227 α 227·03	Th 228 α 228·03	Th 229 α 229·03	Th 230 α 230·03	Th 231 β^-	Th 232 α 232·04	Th 233 β^-	Th 234 β^-	Th 235 β^-			
89	Ac 215 α	Ac 216 α	Ac 217 α	Ac 218 α	Ac 219 α	Ac 220 α	Ac 221 α 221·02	Ac 222 α 222·02	Ac 223 α 223·02	Ac 224 EC, α 224·02	Ac 225 α 225·02	Ac 226 β^-,EC	Ac 227 β^- 227·03	Ac 228 β^-	Ac 229 β^-	Ac 230 β^-	Ac 231 β^-						
88	Ra 214 α 214·00	Ra 215 α 215·00	Ra 216 α 216·00	Ra 217 α 217·01	Ra 218 α 218·00	Ra 219 α 219·01	Ra 220 α 220·01	Ra 221 α 221·01	Ra 222 α 222·02	Ra 223 α 223·02	Ra 224 α 224·02	Ra 225 β^-	Ra 226 α 226·03	Ra 227 β^-	Ra 228 β^-	Ra 229 β^-	Ra 230 β^-						
87	Fr 213 α 213·00	Fr 214 α 214·00	Fr 215 α 215·00	Fr 216 α 216·00	Fr 217 α 217·01	Fr 218 α 218·01	Fr 219 α 219·01	Fr 220 α 220·01	Fr 221 α 221·01	Fr 222 β^-	Fr 223 β^- 223·02	Fr 224 β^-	Fr 225 β^-	Fr 226 β^-	Fr 227 β^-								
86	Rn 212 α 212·99	Rn 213 α 213·99	Rn 214 α 214·00	Rn 215 α 215·00	Rn 216 α 216·00	Rn 217 α 217·00	Rn 218 α 218·01	Rn 219 α 219·01	Rn 220 α 220·01	Rn 221 β^- α	Rn 222 α 222·02	Rn 223 β^-	Rn 224 β^-	Rn 225 β^-	Rn 226 β^-								
85	At 211 EC,α	At 212 α 211·99	At 213 α 212·99	At 214 α 214·00	At 215 α 215·00	At 216 α 216·00	At 217 α 217·00	At 218 α 218·01	At 219 α 219·01														
84	Po 210 α 209·98	Po 211 α 210·99	Po 212 α 212·99	Po 213 α 213·99	Po 214 α 214·00	Po 215 α 215·00	Po 216 α 216·00	Po 217 α 217·01	Po 218 α 218·01														
83	Bi 209 208·98	Bi 210 β^-	Bi 211 α 210·99	Bi 212 β^-	Bi 213 β^-	Bi 214 β^-	Bi 215 β^-																
82	Pb 208 207·98	Pb 209 β^-	Pb 210 β^-	Pb 211 β^-	Pb 212 β^-	Pb 213 β^-	Pb 214 β^-																

Z ↑ N →

Key: U / 233 / α / 233·04 — Chemical symbol; Atomic mass number; Principal decay mode of ground state; Mass (carbon-12 scale)

Heavy box: Stable or sufficiently long-lived to either occur naturally or to be of special usefulness

Half-lives and binding energies in Table 2.1

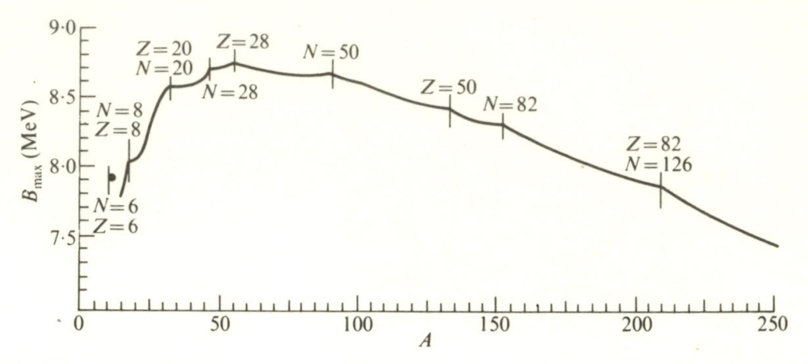

FIG. 2.3. Binding energy per particle for the most stable nuclei as a function of mass number A.

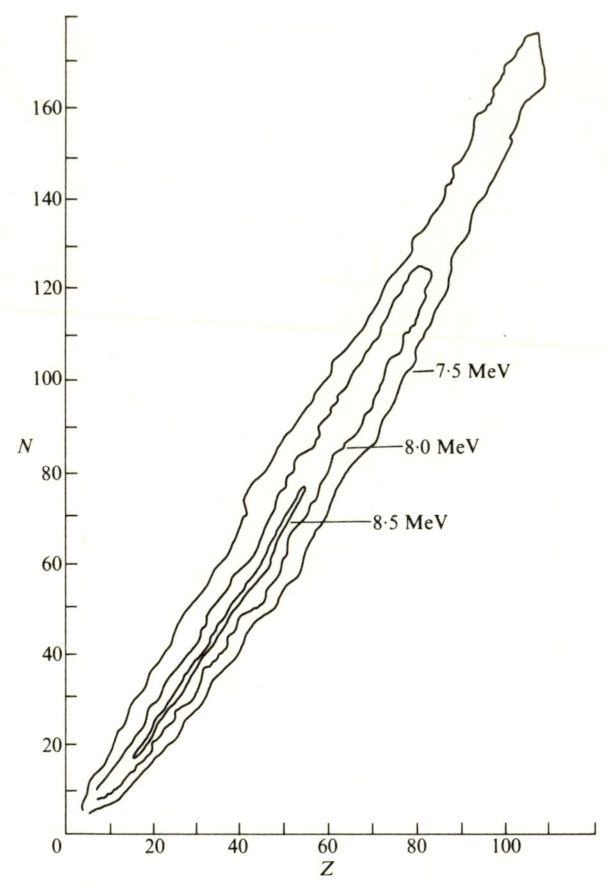

FIG. 2.4. Energy contours for binding energies per particle in the plane of neutron number N and proton number Z.

TABLE 2.1

Binding energies, half-lives, and principal decay modes of the ground states of some heavy nuclei†

Nucleus	Binding energy (MeV)	Half-life	Principal decay
^{210}Po	1645	138·4 days	α
^{211}Po	1650	25 s	α
^{212}Po	1656	45 s	α
^{213}Po	1660	$4{\cdot}2 \times 10^{-6}$ s	α
^{214}Po	1666	$1{\cdot}64 \times 10^{-4}$ s	α
^{215}Po	1670	$1{\cdot}78 \times 10^{-3}$ s	α
^{216}Po	1676	0·15 s	α
^{217}Po	1680	<10 s	α
^{218}Po	1686	3·05 min	α
^{211}At	1648	7·21 hours	Ec, α
^{212}At	1653	0·3 s	α
^{213}At	1659	short	α
^{214}At	1664	2×10^{-6} s	α
^{215}At	1670	$\sim 10^{14}$ s	α
^{216}At	1675	3×10^{-4} s	α
^{217}At	1681	0·03 s	α
^{218}At	1685	2·0 s	α
^{219}At	1690	0·9 min	α
^{215}Rn	1669	10^{-6} s	α
^{216}Rn	1676	$4{\cdot}5 \times 10^{-5}$ s	α
^{217}Rn	1681	$5{\cdot}4 \times 10^{-4}$ s	α
^{218}Rn	1687	0·04 s	α
^{219}Rn	1692	4 s	α
^{220}Rn	1698	55 s	α
^{221}Rn	1702	25 min	β^-
^{222}Rn	1708	3·82 days	α
^{217}Fr	1679	short	α
^{218}Fr	1684	5×10^{-3} s	α
^{219}Fr	1691	0·02 s	α
^{220}Fr	1696	27·5 s	α
^{221}Fr	1703	4·8 min	α
^{222}Fr	1707	14·8 min	β^-
^{223}Fr	1713	22 min	β^-
^{224}Fr	1718	< 2 min	β^-
^{219}Ra	1689	10^{-3} s	α
^{220}Ra	1697	0·02 s	α
^{221}Ra	1702	29 s	α
^{222}Ra	1709	37·5 s	α
^{223}Ra	1714	11·4 days	α
^{224}Ra	1720	3·64 days	α
^{225}Ra	1725	14·8 days	β^-
^{226}Ra	1732	$\sim$ 1600 years	α

Nucleus	Binding energy (MeV)	Half-life	Principal decay
^{227}Ra	1736	41·2 min	β^-
^{228}Ra	1742	6·7 years	β^-
^{221}Ac	1700	short	α
^{222}Ac	1706	5 s	α
^{223}Ac	1712	2·2 min	α
^{224}Ac	1718	2·9 hours	Ec
^{225}Ac	1725	10 days	α
^{226}Ac	1730	29 hours	β^-, Ec
^{227}Ac	1737	21·5 years	β^-
^{228}Ac	1742	6·13 hours	β^-
^{229}Ac	1748	66 min	β^-
^{230}Ac	1753	< 1 min	β^-
^{223}Th	1710	0·9 s	α
^{224}Th	1718	1·05 s	α
^{225}Th	1723	8·0 min	α
^{226}Th	1731	30·9 min	α
^{227}Th	1736	18·2 days	α
^{228}Th	1743	1·91 years	α
^{229}Th	1748	7340 years	α
^{230}Th	1755	8×10^{4} years	α
^{231}Th	1760	25·52 hours	β^-
^{232}Th	1767	$1{\cdot}41 \times 10^{10}$ years	α
^{233}Th	1772	22·4 min	β^-
^{234}Th	1778	24·10 days	β^-
^{225}Pa	1720	0·8 s	α
^{226}Pa	1727	1·8 min	α, Ec
^{227}Pa	1734	38·3 min	α, Ec
^{228}Pa	1740	22 hours	Ec
^{229}Pa	1747	1·5 days	Ec
^{230}Pa	1753	17·5 days	Ec, β^-
^{231}Pa	1760	$3{\cdot}25 \times 10^{4}$ years	α
^{232}Pa	1765	1·31 days	β^-
^{233}Pa	1772	27·0 days	β^-
^{234}Pa	1777	6·75 hours	β^-
^{235}Pa	1783	23·7 min	β^-
^{236}Pa	1788	12 min	β^-
^{237}Pa	1794	39 min	β^-
^{228}U	1739	9·1 min	α
^{229}U	1745	58 min	Ec, α
^{230}U	1753	20·8 days	α
^{231}U	1759	4·3 days	Ec
^{232}U	1766	72 years	α
^{233}U	1772	$1{\cdot}62 \times 10^{5}$ years	α

†Where more than one decay mode has a greater than 10 per cent intensity, the decays are listed in order of probability.

Table 2.1 continued

Nucleus	Binding energy (MeV)	Half-life	Principal decay
^{234}U	1779	2·47 × 10^{5} years	α
^{235}U	1784	7·1 × 10^{8} years	α
^{236}U	1790	2·39 × 10^{7} years	α
^{237}U	1796	6·7 days	β^-
^{238}U	1802	4·51 × 10^{9} years	α
^{239}U	1807	23·5 min	β^-
^{240}U	1812	14·1 hours	β^-
^{231}Np	1756	50 min	α
^{232}Np	1763	13 min	Ec
^{233}Np	1770	35 min	Ec
^{234}Np	1776	4·4 days	Ec
^{235}Np	1783	410 days	Ec
^{236}Np	1789	22 hours	Ec, β^-
^{237}Np	1795	2·14 × 10^{6} years	α
^{238}Np	1801	2·10 days	β^-
^{239}Np	1807	2·35 days	β^-
^{240}Np	1812	63 min	β^-
^{241}Np	1818	16 min	β^-
^{234}Pu	1775	9 hours	Ec
^{235}Pu	1781	26 min	Ec
^{236}Pu	1788	2·85 years	α
^{237}Pu	1794	45 days	Ec
^{238}Pu	1801	86·4 years	α
^{239}Pu	1807	24 390 years	α
^{240}Pu	1813	6580 years	α
^{241}Pu	1819	13·2 years	β^-
^{242}Pu	1825	3·79 × 10^{5} years	α
^{243}Pu	1830	4·98 hours	β^-
^{244}Pu	1836	7·6 × 10^{7} years	α
^{245}Pu	1841	10·5 hours	β^-
^{246}Pu	1847	10·85 days	β^-
^{237}Am	1792	1·3 hours	Ec
^{238}Am	1798	1·9 hours	Ec
^{239}Am	1805	12·1 hours	Ec
^{240}Am	1811	51 hours	Ec
^{241}Am	1818	458 years	α
^{242}Am	1824	16·01 hours	β^-, Ec
^{243}Am	1830	7·95 × 10^{3} years	α
^{244}Am	1835	10·1 hours	β^-
^{245}Am	1841	2·07 hours	β^-
^{246}Am	1846	25 min	β^-
^{238}Cm	1796	2·5 hours	Ec, α
^{239}Cm	1803	3 hours	Ec
^{240}Cm	1810	26·8 days	α
^{241}Cm	1817	35 days	Ec
^{242}Cm	1823	162·5 days	α
^{243}Cm	1829	32 years	α
^{244}Cm	1836	17·6 years	α
^{245}Cm	1842	9·3 × 10^{3} years	α
^{246}Cm	1848	5·5 × 10^{3} years	α
^{247}Cm	1853	1·6 × 10^{7} years	α
^{248}Cm	1859	4·7 × 10^{5} years	α, SF
^{249}Cm	1864	64 min	β^-
^{243}Bk	1827	4·5 hours	Ec
^{244}Bk	1833	4·4 hours	Ec
^{245}Bk	1840	4·98 days	Ec
^{246}Bk	1846	1·8 days	Ec
^{247}Bk	1852	1·4 × 10^{3} years	α
^{248}Bk	1858	16 hours	β^-, Ec
^{249}Bk	1864	314 days	β^-
^{250}Bk	1869	193·3 min	β^-
^{244}Cf	1831	25 min	α
^{245}Cf	1838	44 min	Ec, α
^{246}Cf	1845	35·7 hours	α
^{247}Cf	1851	2·5 hours	Ec
^{248}Cf	1858	350 days	α
^{249}Cf	1863	360 years	α
^{250}Cf	1870	13·2 years	α
^{251}Cf	1875	800 years	α
^{252}Cf	1881	2·65 years	α
^{253}Cf	1886	17·6 days	β^-
^{246}Es	1841	7·5 min	Ec, α
^{247}Es	1848	5·0 min	Ec
^{248}Es	1854	25 min	Ec
^{249}Es	1861	2 hours	Ec
^{250}Es	1867	8 hours	Ec
^{251}Es	1874	1·5 days	Ec
^{252}Es	1879	140 days	α
^{253}Es	1886	20·5 days	α
^{254}Es	1891	276 days	α
^{248}Fm	1852	0·6 min	α
^{249}Fm	1858	2·5 min	α
^{250}Fm	1866	30 min	α
^{251}Fm	1872	7 hours	Ec
^{252}Fm	1879	22·7 hours	α
^{253}Fm	1885	3 days	Ec, α
^{254}Fm	1891	3·24 hours	α
^{255}Fm	1896	20·1 hours	α
^{255}Md	1895	0:6 hours	Ec, α
^{253}No	1877	100 s	α
^{254}No	1885	55 s	α
^{255}No	1892	180 s	α
^{257}Lw	1902	~ 20 s	α

moment of inertia to deduce some features of the nuclear structure of these nuclei. Table 2.2 indicates how closely the experimental energy levels correspond to the quantal rotator values predicted by eqn. (2.2) in the case of ^{244}Cm. We see that up to, and including, the 6^+ level the agreement is excellent. However, a discrepancy of a few per cent has appeared by the time we reach the 8^+ state, and the discrepancy increases rapidly thereafter. Such results are typical throughout this region.

TABLE 2.2

A comparison of the observed 'rotational' energies in ^{244}Cm compared with the predictions of eqn (2.2)

	E_{exp}	E_{rot}
2+	0·0429	0·0429
4+	0·1423	0·1430
6+	0·296	0·3003
8+	0·502	0·5148

Further evidence supporting our interpretation of the 'rotational' bands comes from a study of the gamma-decay spectrum. The sequence of decays . . . $8^+ \to 6^+ \to 4^+ \to 2^+ \to 0^+$ is characterized by the observation of extremely strong electric quadrupole radiation such as would be expected to arise in large-scale charge-density fluctuations corresponding to collective motion.

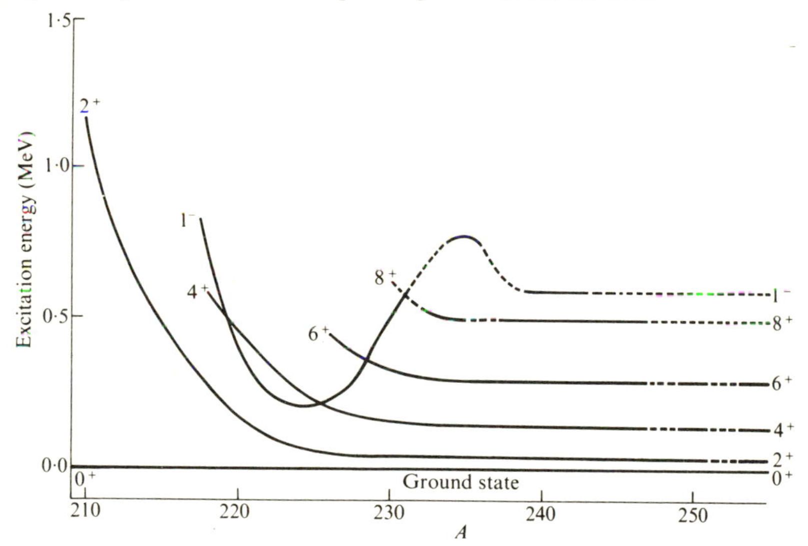

FIG. 2.5. Low-energy excitation spectra of the most stable heavy even–even nuclei as a function of mass number A.

Turning to the spectra of even–even nuclei in the range $210 < A < 230$ the only obvious systematic is that the first excited state is always a 2^+ level and that its energy falls rapidly with increasing mass number. There is always a gap of at least 100 keV between the first 2^+ level and the next excited state, but above this the density of energy levels increases rapidly with excitation energy.

If we study the odd–odd heavy nuclei we find that very little is known about the excitation spectra. Indeed there are fewer than a dozen nuclides throughout the whole region in which confident spin and parity assignments can be made for levels other than the ground state. The only completely unambiguous statement that can be made is that the density of low-lying states is much greater for the odd–odd nuclei than for even–even nuclei. In all cases the first excited state lies within 300 keV of the ground state, and most commonly at less than 100 keV excitation.

Turning to the spectra of odd-A nuclei we consider first the odd–even nuclei, i.e. those with even numbers of protons. In Table 2.3 we present a study of the

TABLE 2.3
Ground-state spins and parities of heavy odd–even nuclei

^{209}Pb	$\frac{9}{2}^+$	^{229}Th	$\frac{5}{2}^+$	^{241}Pu	$\frac{5}{2}^+$
^{211}Pb	$\frac{9}{2}^+$	^{229}U	$\frac{3}{2}^+$	^{241}Cm	$\frac{1}{2}^+$
^{211}Po	$\frac{9}{2}^+$	^{231}Th	$\frac{5}{2}^+$	^{243}Pu	$\frac{7}{2}^+$
^{213}Po	$\frac{9}{2}^+$	^{231}U	$\frac{5}{2}^-$	^{243}Cm	$\frac{5}{2}^+$
^{215}Po	$\frac{9}{2}^+$	^{233}U	$\frac{5}{2}^+$	^{245}Cm	$\frac{7}{2}^+$
^{217}Rn	$\frac{9}{2}^+$	^{235}U	$\frac{7}{2}^-$	^{249}Cf	$\frac{9}{2}^-$
^{223}Ra	$\frac{1}{2}^+$	^{237}U	$\frac{1}{2}^+$	^{251}Cf	$\frac{1}{2}^+$
^{225}Ra	$\frac{3}{2}^-$	^{237}Pu	$\frac{7}{2}^-$	^{253}Cf	$\frac{7}{2}^+$
^{225}Th	$\frac{3}{2}^+$	^{239}Pu	$\frac{1}{2}^+$	^{255}Fm	$\frac{7}{2}^+$
^{227}Th	$\frac{3}{2}^+$	^{239}U	$\frac{5}{2}^+$	^{257}Fm	$\frac{9}{2}^+$

observed ground-state spins of odd–even nuclei. While there is nothing as dramatic as the situation in the even–even nuclei we notice the following regularities:

(1) a large majority of the nuclei have positive-parity ground states;

(2) the negative-parity ground states have relatively high spin;

(3) the nuclei tend to form groups of neighbours with the same ground-state spin.

Table 2.4 shows a study of the observed ground-state spins of even–odd nuclei. Again there is a tendency for neighbouring nuclei to have the same ground-state spin, and the majority of the states have negative parity.

In both odd–even and even–odd nuclei the density of low-lying states is similar to that exhibited by odd–odd nuclei.

As may be seen from Fig. 2.1, among the lighter nuclei the stability of the ground states is completely determined by the energetics of beta decay. Nuclei with too high a neutron–proton ratio emit electrons while those with too low a neutron–proton ratio decay either through the emission of a positron or by capturing an atomic electron. In heavier nuclei electron-capture becomes dominant.

TABLE 2.4
Ground-state spins and parities of heavy even–odd nuclei

Nucleus	J^π	Nucleus	J^π	Nucleus	J^π
^{209}Bi	$\frac{9}{2}^-$	^{227}Pa	$\frac{5}{2}^-$	^{241}Am	$\frac{5}{2}^-$
^{209}At	$\frac{9}{2}^-$	^{229}Pa	$\frac{3}{2}^-$	^{243}Am	$\frac{5}{2}^-$
^{211}Bi	$\frac{9}{2}^-$	^{231}Pa	$\frac{3}{2}^-$	^{243}Bk	$\frac{3}{2}^-$
^{211}At	$\frac{9}{2}^-$	^{233}Pa	$\frac{5}{2}^+$	^{245}Am	$\frac{5}{2}^+$
^{223}Fr	$\frac{3}{2}^+$	^{235}Np	$\frac{5}{2}^+$	^{245}Bk	$\frac{3}{2}^-$
^{223}Ac	$\frac{5}{2}^-$	^{237}Np	$\frac{5}{2}^+$	^{249}Bk	$\frac{7}{2}^+$
^{225}Ac	$\frac{3}{2}^+$	^{239}Np	$\frac{5}{2}^-$	^{253}Es	$\frac{7}{2}^+$
^{227}Ac	$\frac{3}{2}^+$	^{239}Am	$\frac{5}{2}^-$		

In the region of heavy elements there are nuclei which are beta-stable and others which beta-decay. However, because of the existence of alternative decay modes there is a scarcity of naturally occurring beta-stable nuclei in the heavy-mass region. In table 2.5 are listed the heavy beta-stable nuclei. We note that for the odd-A nuclei, with the exception of $A = 247$, there is only a single beta-stable isotope for each mass number. On the other hand, for even-A nuclei there are most commonly two stable isobars and there are no beta-stable odd–odd isobars.

The principal reason that few naturally occurring beta-stable nuclei are observed is the growing importance of the alpha-decay mode in heavy nuclei. The systematics of the alpha-decay energy with mass number are presented in Figs. 2.6 and 2.7. Fig. 2.6 shows the profile of alpha-decay energy versus mass number for the nuclei of maximum beta stability. We see that at first the alpha energy falls sharply to a minimum at $A \sim 240$ and then begins to rise again, although more slowly. This profile should be set against the pattern of generally rising alpha energies resulting from the steadily increasing Coulomb effects. As a guide in Fig. 2.6 two lines have been drawn indicating half-lives of 1 hour and 10^8 years. Nuclei with lower alpha-decay energies, i.e. half-lives longer than 10^8 years, are likely to beta-decay rather than alpha-decay. Nuclei with higher alpha-decay energies, i.e. half-lives shorter than 1 hour, are likely to alpha-decay rather than beta-decay. This leaves a band of nuclei with intermediate alpha energies for which spontaneous fission becomes a practical possibility.

The alpha-decay energies for isotopes of the heavy elements are indicated in Fig. 2.7, and we see that for a given element there is generally a strong decrease

TABLE 2.5
Beta-stable heavy nuclei

A							
208	Pb	222	Ra	236	U, Pu	250	Cm, Cf
209	Bi	223	Ra	237	Np	251	Cf
210	Po	224	Ra, Th	238	U, Pu	252	Cf, Fm
211	Po	225	Ac	239	Pu	253	Es
212	Po, Rn	226	Ra, Th	240	Pu Cm	254	Cf, Fm
213	Po	227	Th	241	Am	255	Fm
214	Po, Rn	228	Th	242	Pu Cm	256	Fm
215	At	229	Th	243	Am	257	Fm
216	Po, Rn	230	Th, U	244	Pu Cm	258	Fm No
217	Rn	231	Pa	245	Cm	259	Md
218	Rn, Ra	232	Th, U	246	Cm, Cf	260	Fm, No
219	Fr	233	U	247	Cm, Bk	261	No
220	Rn, Ra	234	U	248	Cm, Cf	262	No
221	Ra	235	U	249	Cf	263	No

in the alpha-decay energy with increasing mass number. For nuclides with $Z \gtrsim 94$ kinks appear in the alpha-energy systematics.

Perhaps the most striking systematic feature of alpha decay in the heavy nuclei is still the linear relationship between the range that the emitted alpha

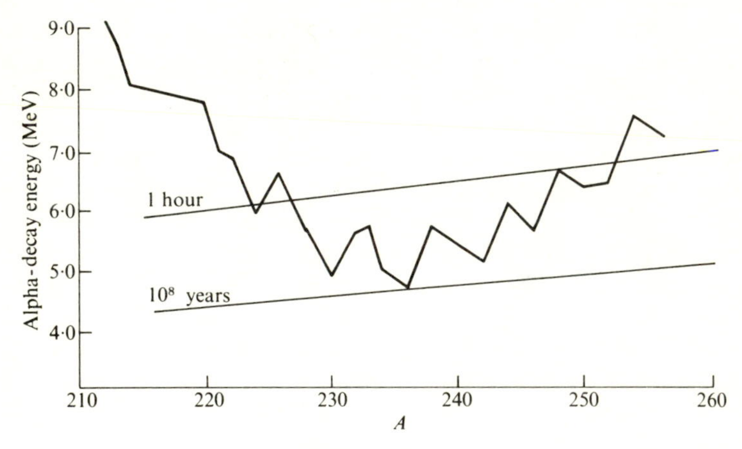

FIG. 2.6. Alpha-decay energies for nuclei of maximum beta stability as a function of mass number A for heavy nuclei.

particles travel in a nuclear emulsion and the logarithm of the alpha-decay half-life, discovered by Geiger and Nuttall in 1911 and illustrated in Fig. 2.8. Since the range is approximately proportional to $E^{1.5}$ the Geiger–Nuttall plot is equivalent to the relationship between half-life and decay energy illustrated in Fig. 2.9.

Turning to spontaneous fission, we see that it is extremely uncommon as the principal decay mode. Table 2.6 shows the nuclei which have a greater than 1 per

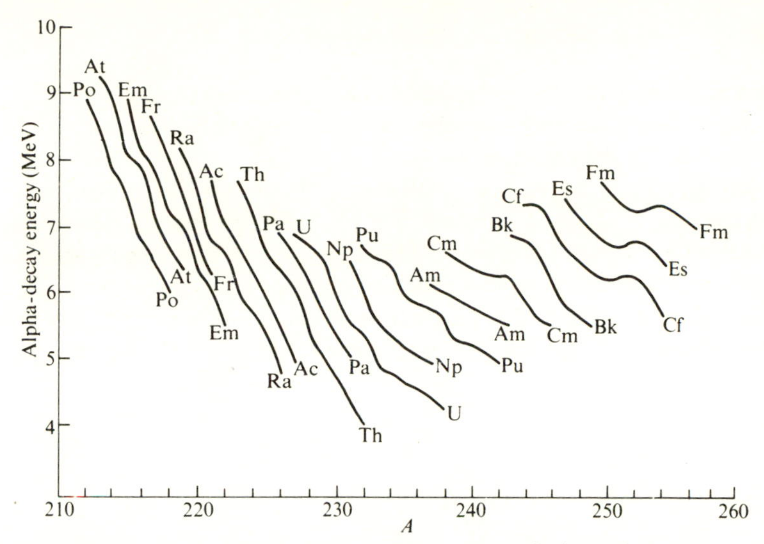

FIG. 2.7. Systematics of alpha-decay energies for heavy isotopes.

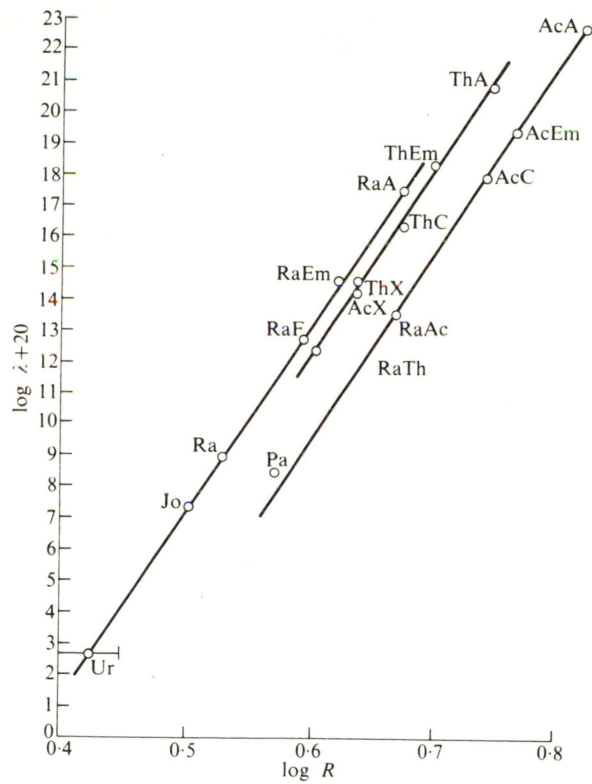

FIG. 2.8. The original plot by Geiger and Nuttall of the relation between the alpha-particle decay constant and the range (From Rutherford, Chadwick, and Ellis, 1930).

cent probability of decaying via spontaneous fission. We note:

(1) the spontaneously fissioning nuclei are extremely massive $A \gtrsim 250$;

(2) they are all even–even nuclei.

In Fig. 2.10 are plotted the spontaneous fission half-lives of the heavy nuclei as a function of Z^2/A. A comparison of Fig. 2.10 and Fig. 2.6 immediately explains the rarity of spontaneous fission. Indeed, so rare is spontaneous fission that very few detailed studies of the fission yields are available. In Fig. 2.11 we plot the spontaneous fission yield versus mass number for ^{252}Cf. This shows a characteristic two-humped yield symmetric about $A \sim 124$. A curve symmetric

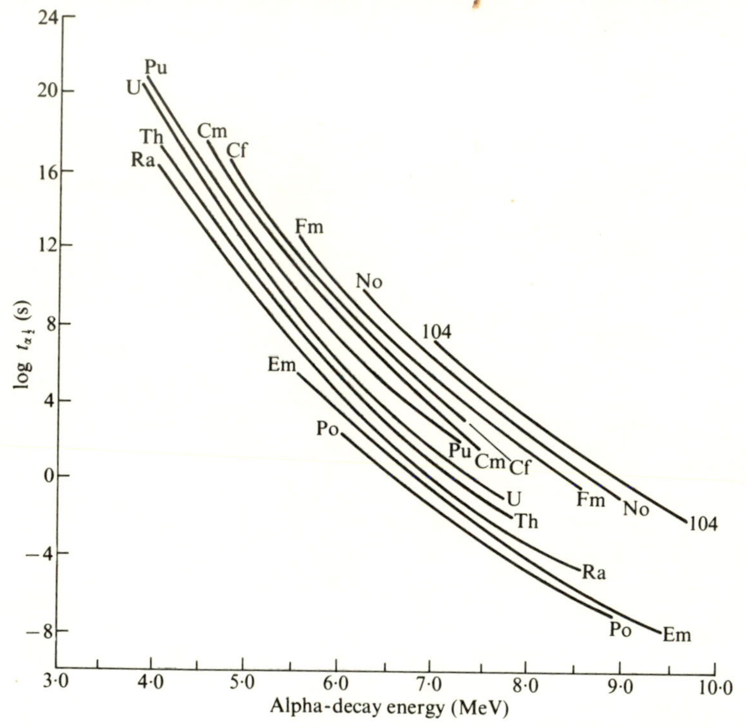

FIG. 2.9 Smooth curves drawn through the experimentally observed alpha-decay half-lives versus alpha-decay energy for various elements.

TABLE 2.6

Nuclei with greater than 1 per cent probability of spontaneous fission

	Percentage		Percentage
^{248}Cm	11	^{250}Cm	~ 100
^{252}Cf	3·1	^{254}Cf	> 90
^{256}Fm	97	^{258}Fm	~ 100
^{252}No	~ 30	^{254}No	
^{260}104			

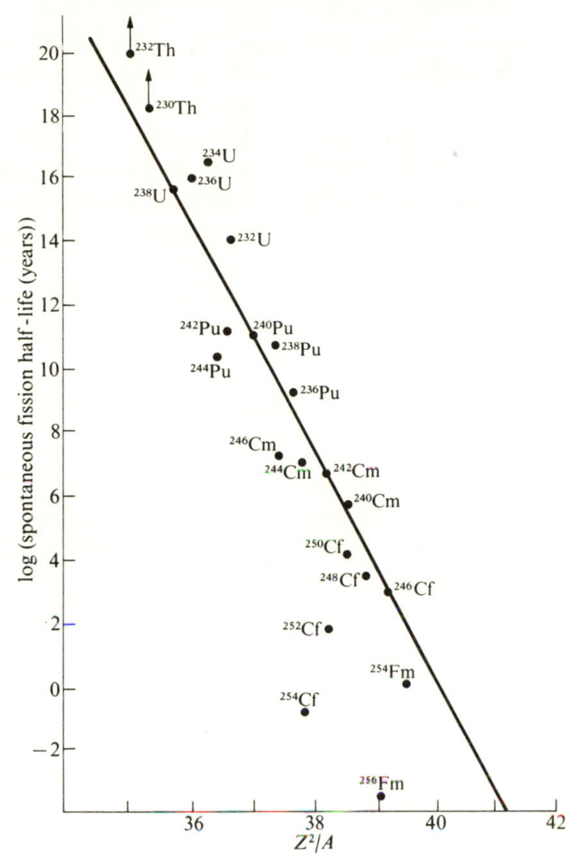

FIG. 2.10. Spontaneous fission half-lives versus Z^2/A.

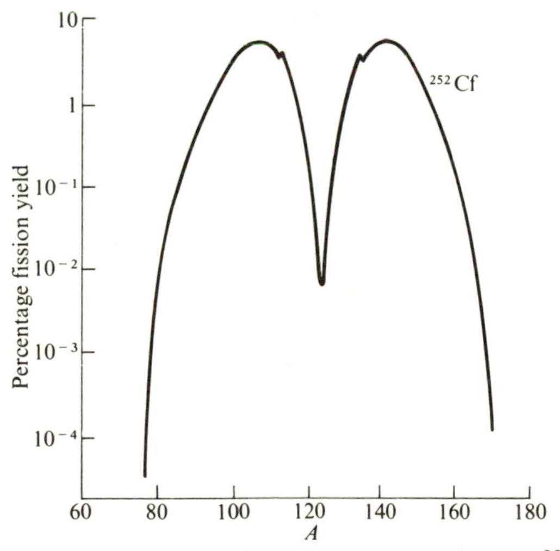

FIG. 2.11. Spontaneous fission yield versus fragment mass for ^{252}Cf.

about $A \sim 124$ indicates a mass defect of the fragments relative to the parent nucleus of ~ 6, and a detailed examination of the fission fragments shows that this is largely accounted for by the emission of neutrons, although emitted alpha particles are not unknown. If we go to parents of heavier mass then the whole yield curve moves to higher values of A with some filling in of the valley between the peak yields. On the other hand, spontaneous fission yields from lighter parents are moved to lower A, and the separation between the peaks is enhanced. In the two cases the average neutron yield is enhanced and diminished respectively.

While spontaneous fission is somewhat of a rarity even amongst the very heavy nuclei, induced fission of one form or another is extremely common. Of particular interest is neutron-induced fission. In Fig. 2.12 we indicate the variation in the neutron-induced fission cross-section for two isotopes of uranium. We note

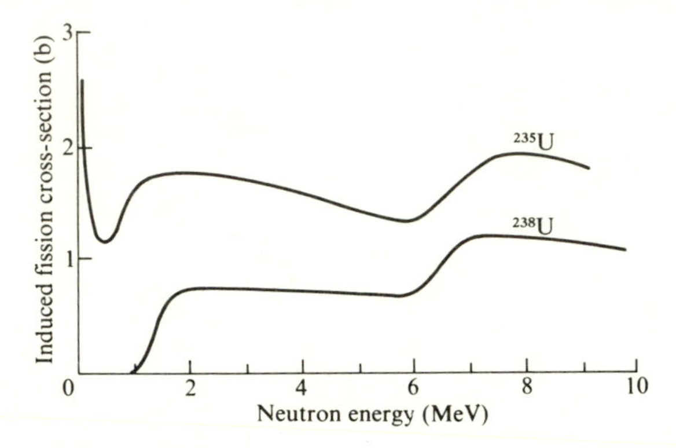

FIG. 2.12. Neutron-induced fission cross-sections versus neutron energy for two isotopes of uranium.

that for ^{238}U the cross-section is essentially negligible until the neutron energy reaches a threshold value of 1–2 MeV. However, for ^{235}U the cross-section is extremely large for low-energy neutrons (thermal neutrons) and falls to a minimum at neutron energies ~ 1 MeV, and thereafter it follows the same general shape as the cross-section for ^{238}U. This difference in behaviour of neutron-induced fission cross-sections for even- and odd-mass isotopes is universal. When a neutron is incident on a nucleus and a compound state is formed, the excited compound nucleus will in general have a choice of decay modes. If we consider the ratio of the cross-section for fission to that for gamma decay as a function of the difference between the neutron binding energy and the fission activation energy we obtain the interesting correlation shown in Fig. 2.13 which is extremely useful for predicting fission cross-sections that are at the limit of observation, e.g. ^{231}Pa.

If we look at the thermal neutron-induced fission yields as a function of mass number we find the following interesting systematics (see Fig. 2.14):

(1) for $A \lesssim 220$ we have a single-humped symmetric distribution about $A \simeq 110$;

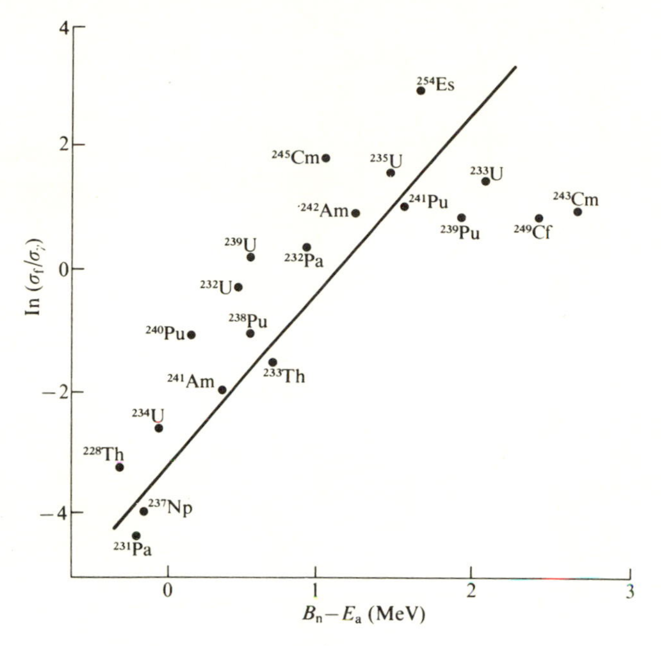

FIG. 2.13. A plot of the ratio of the spontaneous fission cross-section to the gamma-decay cross-section versus the difference between the neutron binding energy and fission activation energy for heavy nuclei.

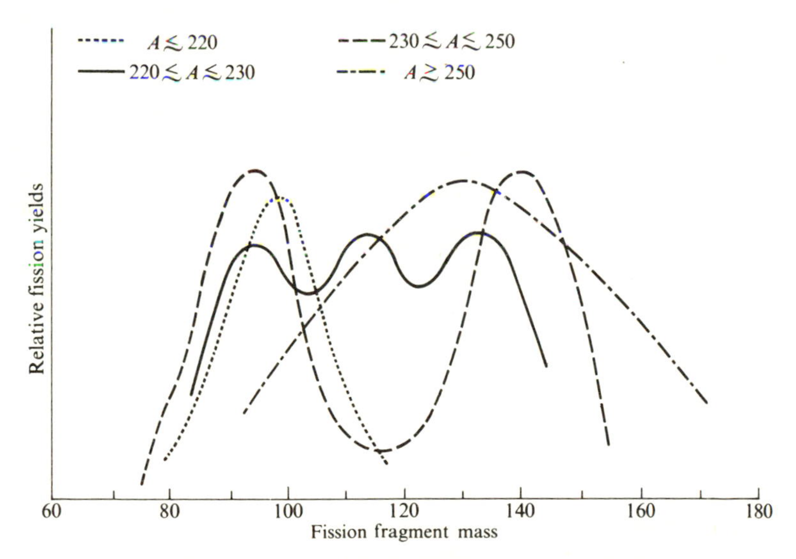

FIG. 2.14. Thermal neutron-induced fission yields versus fragment mass for various mass ranges of heavy nuclei.

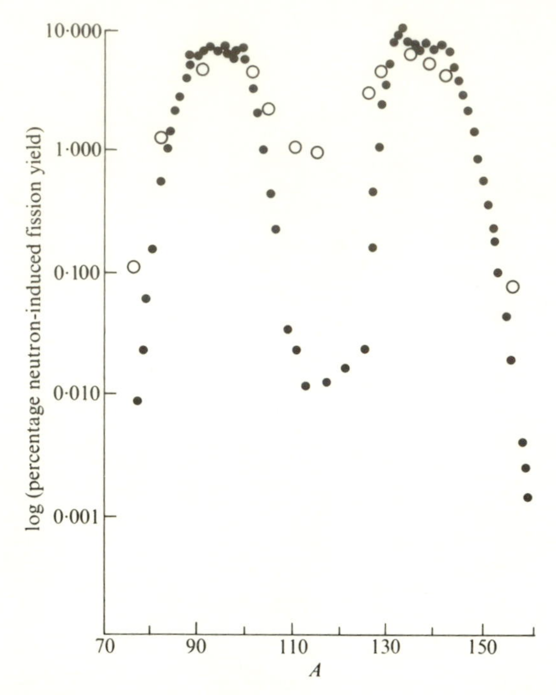

FIG. 2.15. Neutron-induced fission yields versus fragment mass for ^{235}U. The solid points correspond to thermal neutrons while the open circles represent the yields from neutrons at 14 MeV.

(2) for $224 \lesssim A \lesssim 228$ we have a triple-humped distribution symmetric about $A \simeq 110$;

(3) for $230 \lesssim A \lesssim 250$ we have a double-humped yield symmetric about $A \simeq 115$;

(4) for $A \gtrsim 250$ we have a single-humped yield symmetric about $A \simeq 130$.

It is also interesting to study the neutron-induced fission yield as a function of neutron energy. The situation is illustrated for ^{235}U in Fig. 2.15, where we note that as the neutron energy increases the distribution remains symmetric but the valley between the yield peaks is slowly filled in. Again this behaviour is typical of nuclei throughout the region.

Fission may also be induced by gamma-rays. In Fig. 2.16 we plot the measured fission cross-sections versus photon energies for a number of nuclei. We see in all cases that there is a threshold for the photo-fission reaction of ~ 5 MeV for both odd and even isotopes. In the energy range > 5 MeV there is no systematic difference between the even- and odd-mass isotope neutron-induced fission characteristics; the same is true for photo-fission. The photo-fission mass yield distribution is similar to that for neutron-induced fission; the photo-fission mass yield of ^{238}U as a function of photon energy can be seen in Fig. 2.17.

A variation on photo-fission is provided by Coulomb-induced fission. If a

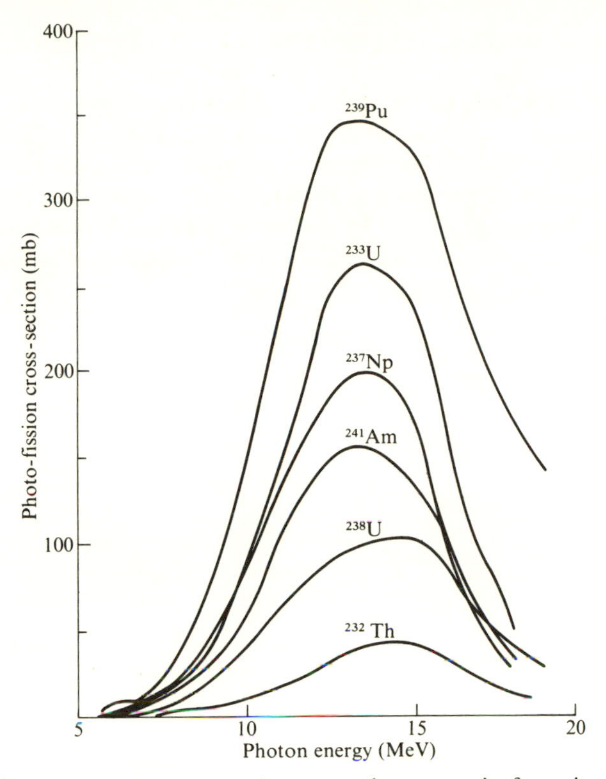

FIG. 2.16. Photo-fission cross-section versus photon energies for various isotopes.

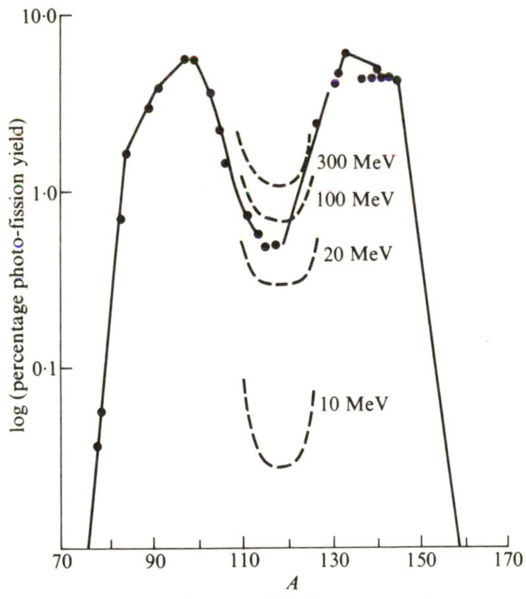

FIG. 2.17. Photo-fission mass yields from ^{238}U for various photon energies. The points represent the yield from 48 MeV photons.

heavy ion is fired at a fissile target it will encounter an extremely large Coulomb barrier. At intermediate ion energies the ion may not be able to approach closely enough the target for nuclear interactions to occur but yet may subject the target to a varying Coulomb energy of several MeV. Such experiments are only practicable with the new generation of heavy-ion accelerators now becoming available. Further information is available in Hyde, Perlman, and Seaborg 1971 and Nuclear Data B1966, 1969, 1970, 1971.

3

NUCLEAR MODELS

3.1. Introduction

In this chapter we shall briefly review the various models which have been developed in attempts to explain nuclear structure, concentrating on those models which appear to have a particular relevance for heavy nuclei (Brown 1967; Davidson 1968; De Shalit and Talmi 1963; Irvine 1972).

We shall not discuss in any detail here the microscopic theory of the nuclear many-body system which provides a justification of the models, although in later chapters we shall have to touch on such problems. In the present chapter we shall adopt an altogether more cavalier and phenomenological approach to the problem. However, I am reminded of a research student I once overheard muttering to himself 'the nucleus is a horrid, nasty, many-body system', and indeed if we hope to explain in any detail the properties of a particular nucleus we must expect to become embroiled in these difficulties.

In Chapter 2 we have seen that there are several dominant systematic features observed in heavy nuclei, and as a first step it must be these that we seek to explain.

3.2. The liquid-drop model

3.2.1. The semi-empirical mass formulae

Our first concern must be to explain in a general fashion the over-all trend of observed masses. If we consider the nucleus to be composed of A nucleons interacting with each other through purely two-body forces then we might predict that the binding energy of the nucleus would vary as the number of pairs of nucleons times an average effective two-body interaction energy, i.e.

$$B(A,Z) \sim \tfrac{1}{2} A(A-1)\bar{V} \tag{3.1}$$

It is quite clear from Fig. 2.3 that this is in fact very different from the observed binding energies. Eqn (3.1) would predict that $\bar{B}/A$ should rise linearly with mass number for all A, whereas we see from Fig. 2.3 that the measured binding energy per particle rises from H to ^{56}Fe (although in a far from linear fashion) and from there on is a decreasing function of A.

Electron-scattering experiments indicate that the charge radius of a nucleus increases with the mass number according to

$$R = r_0 A^{\frac{1}{3}}. \tag{3.2}$$

Scattering experiments with pions indicate that the charge and mass distributions are roughly proportional. Eqn (3.2) is of the form we would expect for the close packing of hard spheres, and indeed the measured values of r_0 are not too dissimilar to the measured radius of the proton-charge form factor at $r_0 \simeq 1{\cdot}2$ fm. We deduce from eqn (3.2) that the nuclear material is essentially incompressible over the mass range of the observed nuclides. In the region of heavy nuclei the radius varies from 7 fm to 8 fm. On the other hand, the range of the nucleon–nucleon interaction is only 2–3 fm. Hence at any instant any particular nucleon could only interact with the small fraction of nucleons which lay within its sphere of influence, (see Fig. 3.1). Since this would be the same for all nucleons in an infinite incompressible material we are led to expect that the binding energy will vary as

$$B(A,Z) \sim An\bar{V} = a_V A, \tag{3.3}$$

where n is the number of nucleons within the sphere of influence of a given nucleon. For a close packing of spheres $n = 12$. The binding energy of the deuteron is 2·2 MeV while the binding energy of the dineutron is ~ 0 MeV, thus we might expect $a_V \simeq 14$ MeV.

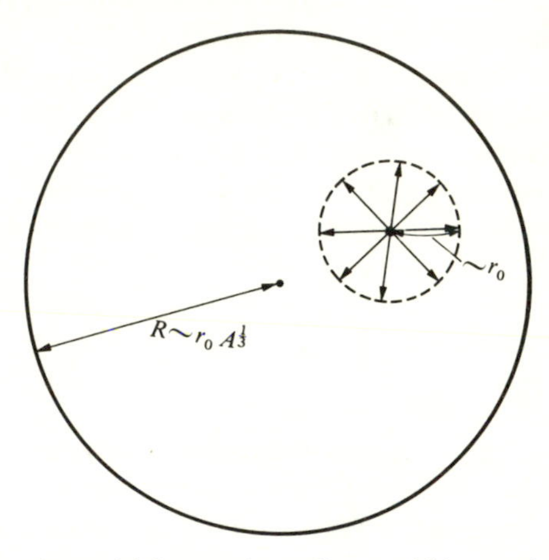

FIG. 3.1. A nucleon with interaction radius r_0 within a nucleus of radius R.

From eqn (3.3) we might expect the binding energy per particle to be constant and this is clearly not in accord with the observations in Fig. 2.3, (p. 8). One obvious omission in our analysis lies in the neglect of nuclear surface effects. Nucleons at the surface of the nucleus will have only incompletely filled spheres of influence (see Fig. 3.2). Thus eqn (3.3) overestimates the binding energy. The number of nucleons in the surface is clearly proportional to the surface area of the nucleus, and using eqn (3.2) we see that eqn (3.3) overestimates the binding energy by a factor

$$E_S = a_S A^{\frac{2}{3}}. \tag{3.4}$$

Our arguments to date would suggest that the binding energy depends only on the mass number A and is independent of the proportion of neutrons to protons. This is clearly not the situation in practice, according to Fig. 2.4. This leads to consideration of another omission – the Coulomb force. This is a long-range force and hence gives rise to a contribution of the form eqn (3.1). However, the Coulomb force acts only between the protons. Also we know that the Coulomb

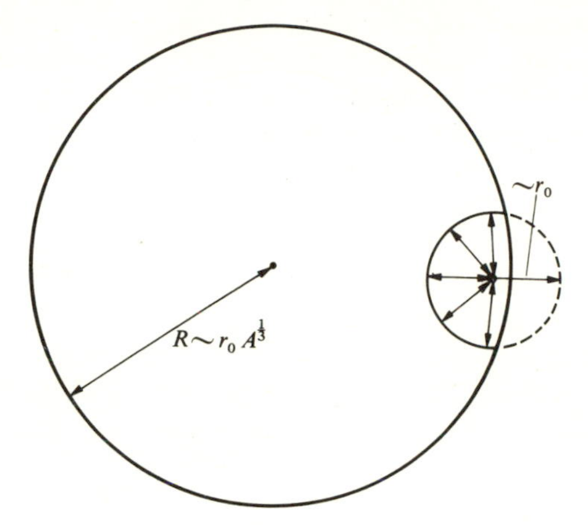

FIG. 3.2. A nucleon at the surface of a nucleus.

self-energy of a uniformly charged sphere is inversely proportional to the radius. Hence we expect a Coulomb contribution to the binding energy of the form

$$E_C = a_C\, Z(Z-1)/A^{\frac{1}{3}}. \tag{3.5}$$

Collecting eqns (3.3), (3.4), and (3.5) together we obtain

$$B(A,Z) = a_V A + a_S A^{\frac{2}{3}} + a_C\, Z(Z-1)/A^{\frac{1}{3}}, \tag{3.6}$$

where a_V is positive and a_S and a_C are negative.

Eqn (3.6) has been deduced assuming that the nucleus is spherical. One definition of a sphere is that it is the three-dimensional figure which contains a given volume with the minimum surface area. Since the nuclear fluid is incompressible any distortion of the nucleus will lead to an increase in the surface area and hence the surface energy. Hence the surface term will oppose distortions of the nucleus. On the other hand, the sphere is the most compact configuration and hence corresponds to a maximum Coulomb energy. Thus the Coulomb term will favour distortions of the nuclear drop. The equilibrium shape of the nucleus will thus be the result of a balance between the surface term and the Coulomb term.

For any arbitrarily shaped drop the Coulomb energy for a uniform charge distribution may be calculated numerically. The surface energy may be estimated as follows: for a classical liquid drop

$$a_S = -4\pi r_0{}^2 S, \tag{3.7}$$

Where S is the surface tension. If S is known by fitting a_S so as to reproduce the binding energies of spherical nuclei, then the surface energy of an arbitrarily shaped drop is given by S times the surface area of the drop, which can again be calculated numerically.

While eqn (3.6) is capable of explaining the general trends illustrated in Fig. 2.3 it fails in one major respect and that is in predicting the very narrow range of neutron–proton ratios observed in stable nuclides of given mass number A i.e. the narrowness of the peninsula of stability (Fig. 2.1). Indeed, as it stands eqn (3.6) would suggest that the most stable nuclei should be composed solely of neutrons, thereby removing the stability-destroying Coulomb interaction. To explain the presence of protons in the nucleus we must remember that neutrons and protons are different phases of the nucleon. According to simple thermodynamics the condition for equilibrium of a phase mixture is the equality of the chemical potentials in the two phases, i.e. in the case of nucleons the neutron and proton Fermi energies should be the same. In the absence of Coulomb effects this implies $N = Z = \frac{1}{2}A$. The average energy per particle in a Fermi gas is proportional to the Fermi energy E_F, and the Fermi energy is proportional to the density ρ to the 2/3 power. Hence the energy of a Fermi gas of n particles in a volume V may be written

$$E_F \propto n\epsilon_F \propto n^{\frac{5}{3}}/V^{\frac{2}{3}}. \tag{3.8}$$

If we keep the volume constant and change the number of particles to $n + \Delta n$, the change in energy may be written

$$\Delta E_F \propto \frac{\partial}{\partial n}\left(\frac{n^{\frac{5}{3}}}{V^{\frac{2}{3}}}\right) \Delta n + \frac{1}{2}\frac{\partial^2}{\partial n^2}\left(\frac{n^{\frac{5}{3}}}{V^{\frac{2}{3}}}\right) \Delta n^2 + \dots. \tag{3.9}$$

Thus if we start with an even mixture of neutrons and protons and convert a number ΔZ of protons into neutrons at constant volume proportional to A, then, since $\Delta N = -\Delta Z$ and

$$\Delta Z = \tfrac{1}{2}(N-Z), \tag{3.10}$$

where N and Z refer to the final nucleus, we find a total symmetry energy of the form

$$E_I = \Delta E_F(N) + \Delta E_F(Z) = a_I\,(N-Z)^2/A, \tag{3.11}$$

and the stability of a given neutron–proton ratio in a given nucleus results from a balance between the Coulomb and symmetry terms.

There remain two comparatively small-scale features of the binding energy which we have not yet accounted for. These are the pairing effects, i.e. the increased stability of even–even nuclei relative to their odd-mass neighbours which are in turn more stable than their odd–odd neighbours, and the shell effects evident in the increased stability of nuclei with N or Z equal to the magic numbers 2, 8, 20, 28, 40, 50, 82, and 126 (neutrons only).

The former may be accommodated phenomenologically by a pairing energy

$$E_P = \tfrac{1}{2}\left\{(-1)^N + (-1)^Z\right\} a_P\, A^{\frac{1}{2}}, \tag{3.12}$$

and we shall discuss the origin of such an effect in Chapter 4. Similarly, the shell effects can be accommodated in a purely phenomenological fashion by a contribution

$$E_{sh} = a_N \sin N'\pi + a_Z \sin Z'\pi + b_N \sin 2N'\pi + b_Z \sin 2Z'\pi + (C_N + C_Z) \sin N'\pi \sin Z'\pi + d, \tag{3.13}$$

where the indices N and Z correspond to the shells of neutrons and protons respectively and N' and Z' are the corresponding fractional occupation of these shells

$$N' = (N - N_j)/(N_{j+1} - N_j), \quad Z' = (Z - Z_j)/(Z_{j+1} - Z_j), \tag{3.14}$$

where N_j and Z_j are the magic numbers associated with the closure of the jth shell, i.e.

$$N_j < N < N_{j+1} \quad \text{and} \quad Z_j < Z < Z_{j+1}. \tag{3.15}$$

The coefficients of eqn (3.13) obtained by Seeger in a least-squares fit to the binding-energy data are given in Table 3.1.

TABLE 3.1
Seeger's shell-correction coefficients

$\leq Z$ or	$N<$	a	b	c	d
8	20	4·008	−0·428	−1.389	13·51
20	50	−0·508	2·331	−0·463	13·51
50	82	−7·636	0·496	1·950	13·51
82	126	−15·63	−2·284	10·67	13·51
126	184	−27·59	2·660	26·99	13·51

The total binding energy of a nucleus may now be written

$$B(A,Z) = a_V A + a_S A^{\frac{2}{3}} + a_C \frac{Z(Z-1)}{A^{\frac{1}{3}}} + a_I (N-Z)^2/A + a_P \tfrac{1}{2}\{(-1)^N + (-1)^Z\} A^{\frac{1}{2}} + E_{sh}, \tag{3.16}$$

and Table 3.2 shows values of the coefficients of eqn (3.16) suggested by various workers. The coefficients quoted in Table 3.2 are obtained by fitting the observed binding energies of nuclei along the peninsula of stability. The expression (3.16) is capable of reproducing most of the observed binding-energy features.

TABLE 3.2
Sets of semi-empirical mass formula coefficients

a_V	a_S	a_C	a_I	a_P
−15·835	18·33	0·045	23·20	11·2
−16·710	18·50	0·750	25·00	$36(A^{-\frac{3}{4}})$

Note that in the second set the pairing energy is parameterized as $V_p A^{-\frac{3}{4}}$.

It has greatest difficulty fitting the observations in two regions of the periodic table – the rare-earth nuclei and the heavy nuclei. In both regions there is considerable evidence that the nuclei are not spherical, and thus we require corrections to the surface-energy and the Coulomb-energy expressions as discussed above. This requires that we have some detailed information about the shapes of nuclei, and we shall return to discuss this point later in the present chapter. There is an additional complication in the heavy nuclei that the shell energy E_{sh} is not defined since while we know that ^{208}Pb corresponds to a doubly magic nucleus we have not observed the next shell closure. An assumption about the equivalence of neutrons and protons might suggest that element 126 would be magic however the importance of Coulomb effects in this region make such an assumption implausible.

The expression (3.16) has been extrapolated to make predictions regarding nuclei away from the peninsula of stability. First, attempts have been made to extrapolate transversely to regions of extremely proton- or neutron-rich light nuclei. Secondly, attempts have been made to extrapolate along the peninsula to the region of superheavy nuclei. The difficulties mentioned above make these latter extrapolations extremely uncertain, and we shall return to improve upon eqn (3.16) in Chapter 6.

3.2.2. Vibrational states

In the discussion of ground-state binding energies we have assumed that the liquid drop of nuclear matter is either spherical or nearly spherical and that it is static. When we come to consider a model for the excitation spectrum of the nucleus we must consider the possible modes of excitation of the liquid drop. For the low-energy excitations of a spherical liquid drop these are obviously the vibrational modes about the equilibrium spherical configuration.

If we describe the instantaneous shape of the nuclear surface by

$$R(t) = R_0 \left(1 + \sum_{\lambda\mu} \alpha_{\lambda\mu}(t)\, Y_{\lambda\mu}(\theta,\phi)\right) \tag{3.17}$$

we can use the theory of small oscillations applied to the generalized coordinate $\alpha_{\lambda\mu}$ to obtain the effective kinetic energy of vibration

$$T = \tfrac{1}{2} \sum B_\lambda \,|\dot{\alpha}_{\lambda\mu}|^2 \tag{3.18}$$

and the elastic potential energy

$$V = \tfrac{1}{2} \sum C_\lambda \,|\alpha_{\lambda\mu}|^2, \tag{3.19}$$

where B_λ is an effective inertial parameter, which for an irrotational fluid at a uniform density ρ has the form

$$B_\lambda = \rho R_0{}^5/\lambda, \tag{3.20}$$

and the potential parameters C_λ for a uniformly charged irrotational liquid drop are given by

$$C_\lambda = SR_0{}^2(\lambda - 1)(\lambda + 2) - \frac{3}{2\pi} Z^2 e^2 (\lambda - 1)/R_0(2\lambda + 1), \tag{3.21}$$

where S is the surface tension of eqn (3.7) and Ze is the total charge.

We see that, in the small oscillation approximation, the drop executes harmonic vibrations in the generalized coordinates $\alpha_{\lambda\mu}$ with frequencies

$$\omega_\lambda = (C_\lambda/B_\lambda)^{\frac{1}{2}}. \tag{3.22}$$

Assuming that the energies are quantized in the usual manner we obtain the excitation spectrum

$$E_{n\lambda} = n_\lambda \hbar \omega_\lambda, \tag{3.23}$$

where n_λ is the number of phonons in the mode λ. These excitations carry angular momentum λ with z component μ and parity $(-1)^\lambda$. Also, being bosons, only the symmetric states of the multi-quanta system can exist.

As presented the spectrum of states would have integer angular momentum and hence cannot be used to describe odd-A nuclei. Further, the ground state of the system would be spherically symmetric and have the spin assignment $J = 0$. A glance at Table 3.3 shows that there are no odd–odd nuclei in the heavy-nuclei region with angular-momentum-zero ground-state spin assignments (with the possible exception of ^{234}Np). Thus the only candidates for our model are the even–even nuclei.

TABLE 3.3

Ground-state spins and parities of heavy odd–odd nuclei

Nucleus	J^π	Nucleus	J^π
^{210}Bi	1^-	^{210}At	(5^+)
^{212}Bi	1^-		
^{214}Bi	1^-		
^{216}At	1^-		
^{228}Pa	(3^+)		
^{230}Pa	(2^-)		
^{234}Pa	(4^+)	^{234}Np	(0^+)
^{236}Np	(1^-)		
^{238}Np	2^+		
^{242}Am	1^-		
^{244}Am	(6^-)		
^{246}Am	(2^+)		
^{248}Bk	(8^-)		
^{250}Bk	(2^-)		
^{252}Es	(7^+)		
^{254}Es	(7^+)		

From eqns (3.20), (3.21), and (3.22) we see that there is no $\lambda = 0$ mode. This is consistent with our assumption of the incompressibility of nuclear matter since $\lambda = 0$ would correspond to a 'breathing' mode in which the nucleus remains spherical and simply expands and contracts. Similarly, $\omega_1 = 0$ and this is also gratifying since the $\lambda = 1$ distortion corresponds to a movement of the centre of the drop relative to a space-fixed origin, the drop remaining spherical. Since we are only interested in the intrinsic excitations we do not wish to consider the gross motion of the nucleus as a whole.

Thus the first excited state should be a 2^+ state at $\hbar\omega_2$ corresponding to a single $\lambda = 2$ phonon. It is gratifying to note in Fig. 2.5 (p.11.) that, with the sole exception of ^{208}Pb, the first excited state is a 2^+ level in all even–even nuclei throughout the heavy-mass region. At an energy of $2\hbar\omega_2$ we should expect to find a 0^+, 2^+, 4^+ degenerate triplet of states corresponding to the coupling of two $\lambda = 2$ phonons. No such triplet of states is observed anywhere in the heavy-mass region. However, in the range $210 < A < 220$ there appear to be a number of nuclei exhibiting incomplete triplets at approximately the right excitation. We illustrate this with the example of ^{214}Po in Fig. 3.3. A similar situation is found in ^{212}Pb, ^{212}Po, ^{214}Pb, ^{216}Rn, and ^{218}Rn. In all cases the missing member of the

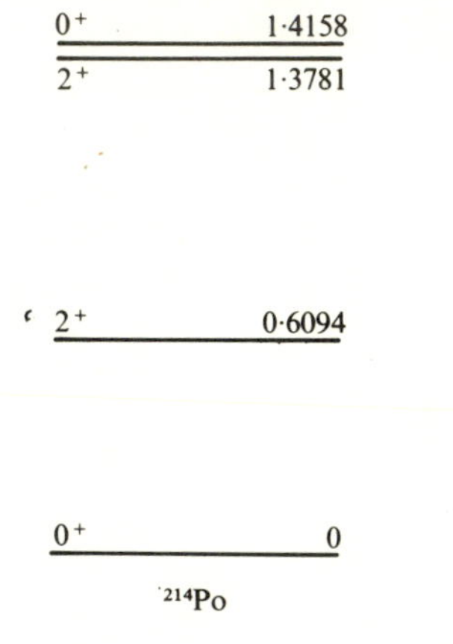

FIG. 3.3 The 'vibrational' spectrum of ^{214}Po.

triplet is the 4^+ level. We should not hastily discard our model simply because of the non-observation of the 4^+ member of the triplet. There are good reasons why such a state may be hard to observe experimentally. Traditionally, the states of the even–even nuclei in this region are populated either by beta decay of the odd–odd neighbouring nuclei or by alpha decay of more massive even–even ground states. This point is illustrated for the case of ^{212}Po in Fig. 3.4. Typically, the energetics are such that the alpha decay only populates the ground state and the first excited state, and hence there is little prospect of seeing the 4^+ level. Commonly the odd–odd ground states have a 1^- spin and parity assignment, and hence beta decay to a 4^+ level would be strongly forbidden. The best prospects for detecting the 4^+ level are probably in heavy-ion alpha-particle transfer reactions. The new generation of heavy-ion accelerators should soon be able to settle this question. 4^+ levels at the correct excitation are not unknown in this region, for example, a low-lying 4^+ level in ^{210}Po is populated by electron-capture in the 5^+ ground state of ^{210}At, while a 4^+ level at 0·47 MeV in ^{220}Ra is populated by the alpha decay of ^{224}Th.

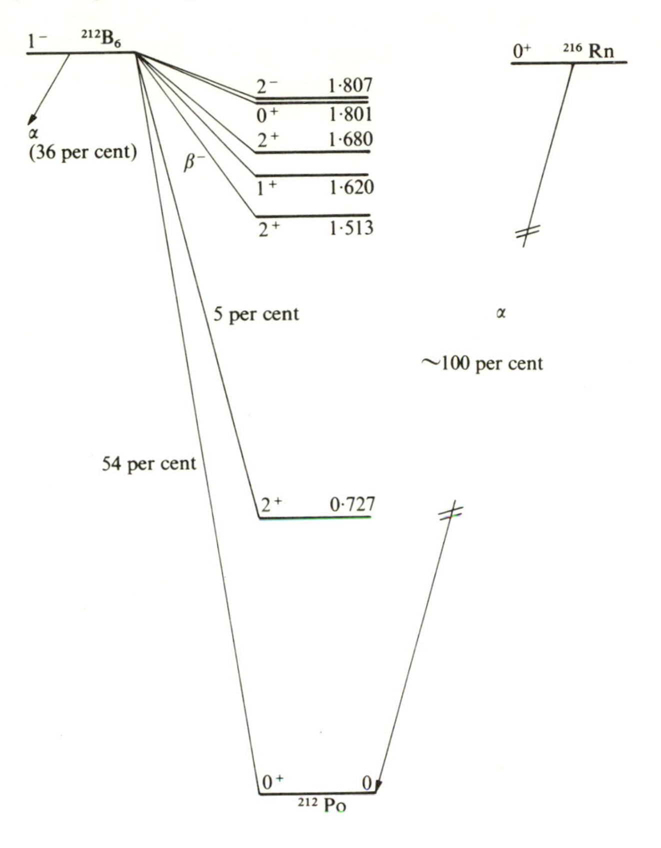

FIG. 3.4. The population of levels in ^{212}Po.

At an energy of $\hbar\omega_3$ we expect to find a 3^- state corresponding to a single $\lambda = 3$ phonon. While there are low-lying 3^- states in the spectra of heavy even–even nuclei only in the case of ^{208}Pb is it the lowest negative-parity state. Elsewhere the first excited negative-parity level has the assignment 1^-.

According to the argument given above a 1^- state cannot arise from the vibrations of a spherical drop except as an oscillation of the centre of mass. However, the nuclear droplet is different from a water droplet in that it is a two-fluid system composed of neutrons and protons. Hence it is possible to have the two fluids oscillating in a 1^- mode out of phase with each other and such that the centre of mass of the whole system is stationary. Since neutrons and protons interact with each other through a predominantly attractive potential this out-of-phase oscillation may be expected to lie at a relatively high energy. In lighter nuclei such excitations at 10–20 MeV are observed in dipole absorption spectra. While it is impossible to observe individual dipole states in the heavy nuclei at 10–20 MeV excitation it is interesting to look back at Fig. 2.16 in which we see that the photo fission cross-section peaks around 15 MeV in exactly the same way that the dipole absorption cross-section does in lighter nuclei. We can then interpret the enhanced photo fission yield as being due to the incident photons exciting a broad resonance in the two-fluid vibrational spectrum. This of course does not explain the low-lying 1^- level observed throughout the heavy-nuclei region (see Fig. 2.5). In the region where there is some evidence of vibrational states ($A \lesssim 220$) the 1^- state lies at a relatively high excitation and is unlikely to have the character of a vibration of a spherical equilibrium shape.

Eqn (3.21) shows clearly the competition for stability between the surface energy and the Coulomb energy. As the Coulomb energy increases the coefficients C_λ and hence the energy of the corresponding phonons decrease. The excitation energy of the quadrupole ($\lambda = 2$) phonon becomes zero when

$$Z^2/A = 40\pi r_0{}^3 S/3e^2. \tag{3.24}$$

Using the parameters of the semi-empirical mass formula of Table 3.2 we find the critical value of Z^2/A lies between 40 and 50. However, since we cannot believe that the nucleus will behave exactly like a classical incompressible irrotational liquid drop we can only take this value of the critical neutron–proton ratio as a rough guide. In Fig. 3.5 the value of Z^2/A versus A is plotted for the most stable even–even nuclei, and we see that it rises steadily as we move through the heavy-mass region. Thus we should not be surprised if we find that a phase transition to

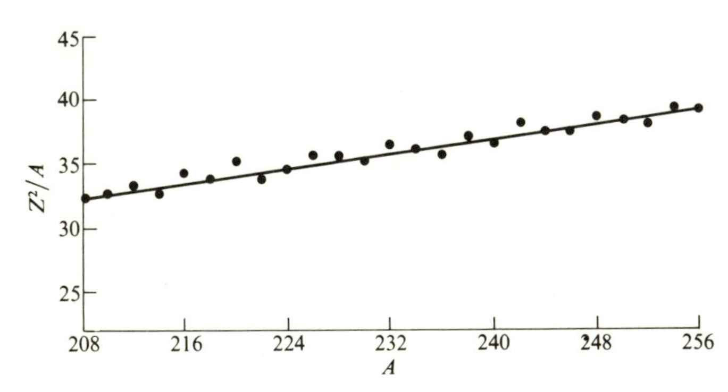

FIG. 3.5. A plot of Z^2/A for the most stable even–even heavy nuclei versus mass number A.

a permanently deformed configuration occurs either in the ground state as we increase the mass or even in the low-energy excitation of a nucleus with a spherical ground state. Such different spatial configurations are called shape isomers.

3.2.3. Rotational states

If the ground state of the nucleus is permanently deformed then we introduce a new mode of excitation, i.e. collective rotations, and if the nucleus rotates as a rigid body this would lead to a spectrum of a ground-state rotational band with

$$E_J = J(J+1)\hbar^2/2\mathscr{I}_{\text{eff}}, \tag{3.25}$$

where $\mathscr{I}_{\text{eff}}$ is the effective moment of inertia. As in the case of the vibrational states the only candidates for such a model based on a $J = 0$ ground state are the even–even nuclei, and indeed, as we have already observed in Fig. 2.5, we have an excellent qualitative agreement for nuclei with $A \gtrsim 230$.

If the deformed shape of the nucleus has some permanence then in a body-fixed reference frame we shall have

$$R = R_0\left\{1 + \sum_{\lambda\mu} a_{\lambda\mu}\, Y_{\lambda\mu}(\theta', \phi')\right\}, \tag{3.26}$$

where the structure coefficients $a_{\lambda\mu}$ are time-independent and the angles θ' and ϕ' are measured relative to the body-fixed frame. This frame is rotating relative to a fixed laboratory frame, and the Euler angles relating the body-fixed frame

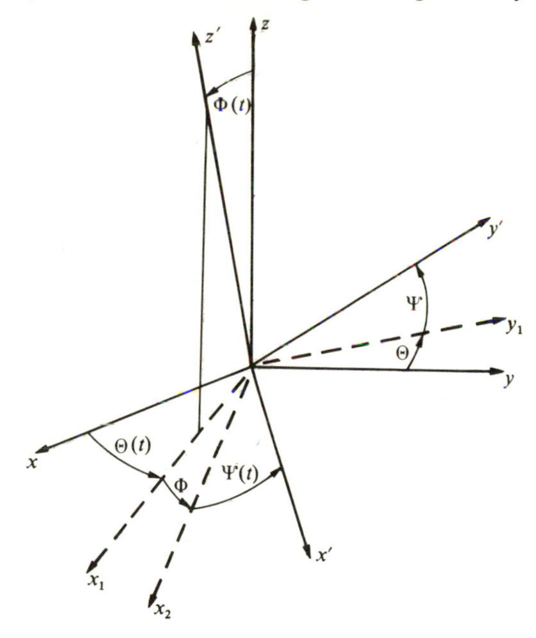

FIG. 3.6. Euler angles.

relative to the laboratory-fixed frame will be time-dependent (see Fig. 3.6). In the laboratory frame eqn (3.26) becomes

$$R(t) = R_0\left\{1 + \sum_{\lambda\mu} \alpha_{\lambda\mu}(t)\, Y_{\lambda\mu}(\theta, \phi)\right\}, \tag{3.27}$$

where, since

$$Y_{\lambda\mu}(\theta,\phi)=\sum_{\mu'}\mathscr{D}^{\lambda}_{\mu\mu'}(\Theta(t),\Phi(t),\Psi(t))\,Y_{\lambda\mu'}(\theta',\phi'),\tag{3.28}$$

we have

$$a_{\lambda\mu}=\sum_{\mu}\mathscr{D}^{\lambda}_{\mu\mu'}(t)\,\alpha_{\lambda\mu}(t).\tag{3.29}$$

If we restrict ourselves to quadrupole deformations ($\lambda = 2$) then there are in general five independent variables $a_{2\mu}$, $\mu = 2, 1, 0, -1, -2$. If the body-fixed axes are chosen to correspond to the principal axes then this reduces to two independent variables, say a_{20} and a_{22}, which together with the Euler angles provide a complete description of the system.

The total Hamiltonian may now be written

$$\mathcal{H}=\tfrac{1}{2}\sum_{K=1}^{3}\mathscr{I}_k\omega_k^2+\tfrac{1}{2}C_2\beta^2,\tag{3.30}$$

where the effective principal moments of inertia are

$$\mathscr{I}_k=4B_2\beta^2\sin^2(\gamma-k(2\pi/3))\quad k=1,2,3,\tag{3.31}$$

with

$$a_{20}=\beta\cos\gamma,\quad a_{22}=\tfrac{1}{2}\beta\sin\gamma.\tag{3.32}$$

If the body-fixed z-axis is an axis of symmetry then

$$\mathscr{I}_1=\mathscr{I}_2=\mathscr{I}_{\text{eff}}=3B_2\beta^2\ \text{ and }\ \mathscr{I}_3=0,\tag{3.33}$$

and using the irrotational value of B_2 given by eqn (3.20) we obtain

$$\mathscr{I}_{\text{eff}}=\frac{45}{16\pi}\mathscr{I}_{\text{rigid}}\,\beta^2,\tag{3.34}$$

where $\mathscr{I}_{\text{rigid}}$ is the rigid-body moment of inertia for a sphere of radius R_0, i.e.

$$\mathscr{I}_{\text{rigid}}=\frac{2}{5}MR_0^2.\tag{3.35}$$

Comparing (3.34) with the experimental result (2.3) we estimate that $\beta \sim 0{\cdot}3$ for nuclei with $230 \lesssim A \lesssim 250$. We note that the effective moment of inertia is zero unless the nucleus is distorted. This is consistent with the quantal result that a spherically symmetric object cannot give rise to rotations. Classically it corresponds to the rotational mode being a tidal flow (see Fig. 3.7) rather than a rigid rotation of the nucleus. We shall return to this and related points in Chapter 5.

If we allow the new equilibrium shape to be distorted, i.e. β and γ are not constants but become time-dependent, then the small oscillation Hamiltonian becomes

$$\mathcal{H} = \tfrac{1}{2}B_2(\dot{\beta} + \beta^2\dot{\gamma}^2) + \tfrac{1}{2}\sum_{k=1}^{3} \mathcal{L}_k{}^2/2\mathcal{I}_k + \tfrac{1}{2}C_2\beta^2, \tag{3.36}$$

and we see the possibility of β and γ vibrations of the nucleus. The β-vibrations conserve the axis of symmetry (see eqn (3.31)) while the γ-vibrations destroy the symmetry.

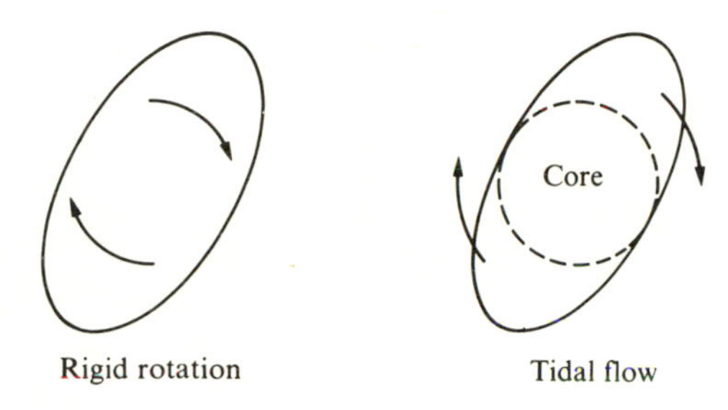

FIG. 3.7. Classical rigid rotation and tidal flow.

If there is an axis of symmetry then there will be three constants of motion, J, M, and K, i.e. the total angular momentum, its component along the space-fixed z-axis, and its component along the axis of symmetry (see Fig. 3.8). If the coupling between the three modes of collective excitation, i.e. rotations and β- and γ-vibrations, is small then we can write the eigenfunctions of the Hamiltonian (3.36) as a product of a β-vibration function $f_{Jn\lambda}(\beta)$, and γ-vibration function $g_K^{Jn}(\gamma)$, and the eigenfunctions for a rigid rotator $D_{MK}^J(\Theta, \Phi, \Psi)$, where the quantum numbers n and λ are the same as in eqn (3.23). Indeed for $\gamma = 0$ the functions $f_{Jn\lambda}(\beta)$ simply describe harmonic oscillations. When there is no axis of symmetry K is no longer a conserved quantity and

$$\Psi_{JM,\,n\lambda} = f_{JM\lambda}(\beta) \sum_{K=1}^{J} g_K^{Jn}(\gamma)\,\mathcal{D}_{MK}^J(\Theta, \Phi, \Psi). \tag{3.37}$$

The functions g_K^{Jn} are restricted by the requirements that the wavefunction be an invariant single-valued function with respect to any transformation which leaves the $\alpha_{2\mu}$s invariant, e.g. a rotation by 180° about the axis of symmetry implies

$$g_K^{Jn}(\gamma) = \mathrm{e}^{\mathrm{i}\pi K}\, g_K^{Jn}(\gamma), \tag{3.38}$$

and hence in order for $g_K^{Jn}(\gamma)$ to be single-valued we have that K must be an even integer. Similarly, a rotation by 180° about the body-fixed x-axis leads to the requirement

$$g_K^{Jn}(\gamma) = e^{i\pi(J+K)} g_{-K}^{Jn}(\gamma), \qquad (3.39)$$

and hence for an axially symmetric nucleus the rotational eigenfunction may be written

$$\Psi_{JMK} = \left\{\frac{2J+1}{16\pi^2(1+\delta_{K_0})}\right\}^{\frac{1}{2}} \{\mathscr{D}^J_{MK} + (-1)^J \mathscr{D}^J_{M-K}\}, \qquad (3.40)$$

and we see that for $K = 0$ only even-J states exist.

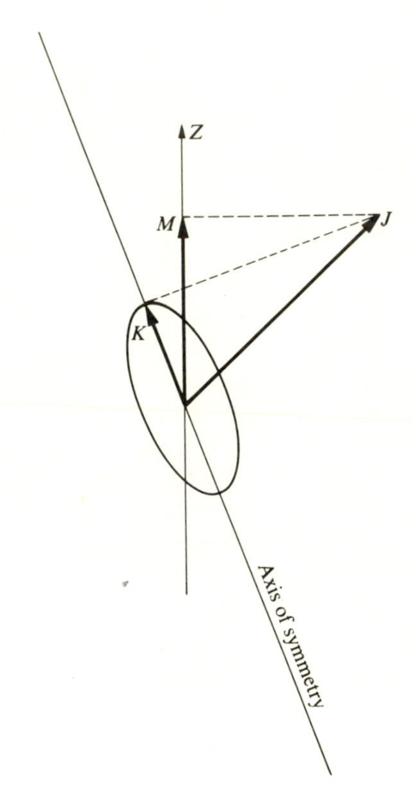

FIG. 3.8. Angular momenta of a deformed nucleus.

We now have an explanation of the spectra of even–even nuclei with $A \gtrsim 230$ shown in Fig. 2.5 in terms of the rotations of an axially symmetric non-spherical nuclear droplet. The low-lying 1^- state corresponds to a β-vibration, i.e. $\lambda = 1$ phonon, coupled to the $J = K = 0$ rotational ground state. Built upon this vibrational state we can again obtain a negative-parity rotational band. At an even higher energy we might expect the re-emergence of the 0^+ spherical configuration which in lighter nuclei corresponds to the ground state. We illustrate these features in Fig. 3.9 with the examples of ^{228}Th and ^{240}Pu.

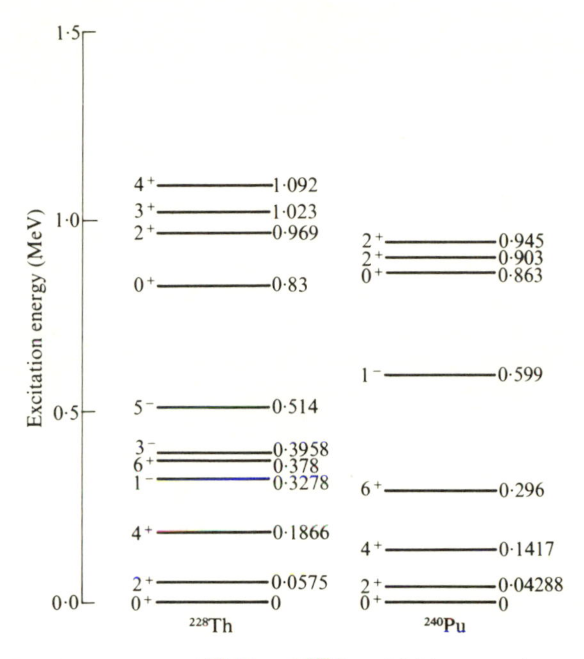

FIG. 3.9. Low-lying spectra of ^{228}Th and ^{240}Pu exhibiting ground-state rotational bands.

The following apparent paradox may be evident to the reader. The ground states of the heavy even–even nuclei have the assignments 0^+ implying that they are spherical. However, we have interpreted the spectra in terms of the rotations of a deformed droplet. The solution of the paradox lies in the uncertainty principle. The angle between the symmetry axis and the space-fixed frame does not commute with the component of angular momentum along the symmetry axis. A measured value of $J = 0$ implies $K = 0$, and since K is now known the orientation of the drop is completely undefined. Hence the quantum state corresponds to a deformed intrinsic configuration averaged over all possible orientations, the result is a spherical ground state.

We are now in a position to review the spectra of the even–even nuclei in the heavy-mass region. Starting with ^{208}Pb we have an extremely stable spherical ground state. The first excited state is a collective octupole ($\lambda = 3$) vibration about the spherical configuration. There is also an out-of-phase collective vibration of neutrons and protons at ~ 15 MeV (the giant dipole resonance). As the mass increases the nuclei become softer against vibrations and the first excited state becomes the quadrupole ($\lambda = 2$) vibration about a spherical ground-state configuration. However, the resistance to distortions falls rapidly with increasing mass, and at modest excitation energies we find a 1^-state corresponding to a vibration about a non-spherical axially symmetric configuration. Between $A = 220$ and $A = 230$ a phase transition occurs to a deformed axially symmetric ground state which supports an even-parity rotational band. The lowest negative-parity state is now always the 1^-vibration based on this non-spherical ground state; this in turn supports a negative-parity rotational band. There is still evidence at higher energies in the spectra of the residue of the spherically symmetric configuration.

Turning to odd-A nuclei, since the ground states of the neighbouring even–even nuclei are stable 0^+ collective configurations, we might expect the spin and parity assignments to be dictated by the states of the odd nucleon. If we take as the prototype odd-A nuclei ^{209}Pb and ^{209}Bi we see that there is some evidence for this amongst the lighter nuclei, i.e. the odd-proton nuclei have the same spin and parity as ^{209}Pb. We shall postpone further discussion of this until after we have treated the individual-particle model.

3.2.4. *Fission* (Cohen and Swiatecki 1962, Nix and Swiatecki 1965, Wheeler 1955)

Restricting ourselves to axially symmetric quadruple deformations

$$R = R_0(1 + \alpha_2 P_2(\cos\theta)) \tag{3.41}$$

we see that as α_2 increases the nuclear shape passes through the series of configurations indicated in Fig. 3.10. At $\alpha_2 = 2$ it closely resembles what we might

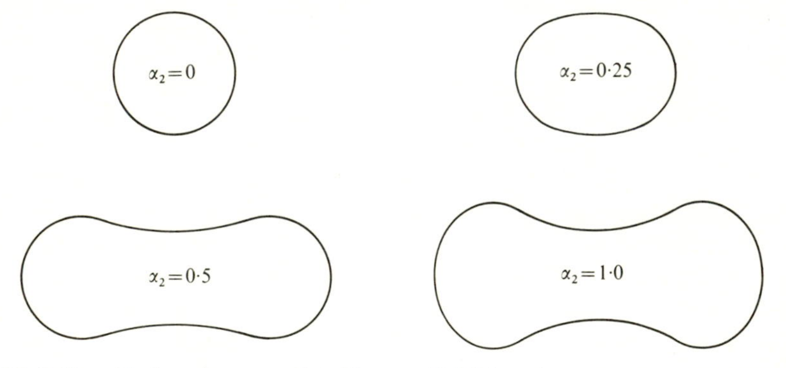

FIG. 3.10. Nuclear shapes predicted by eqn (3.41) (not drawn to represent constant volume).

expect at the scission point for symmetric fission. The condition for stability against quadrupole deformation is still (eqn (3.24))

$$Z^2/A \leqslant 40\pi r_0{}^3 S/3e^2. \tag{3.42}$$

This condition is obtained by equating the Coulomb energy and the surface energy terms arising from the distortion of a liquid drop. From the observed rotational spectra and eqns (3.25) and (3.34) we see that the equilibrium deformation occurs for $\alpha_2 \lesssim 0{\cdot}3$. For such small deformations the surface and Coulomb energies have only changed marginally compared with their values in the spherical configuration. If we now consider the symmetric scission of the drop, then for two spherical touching daughter nuclei (Fig. 3.11), the surface

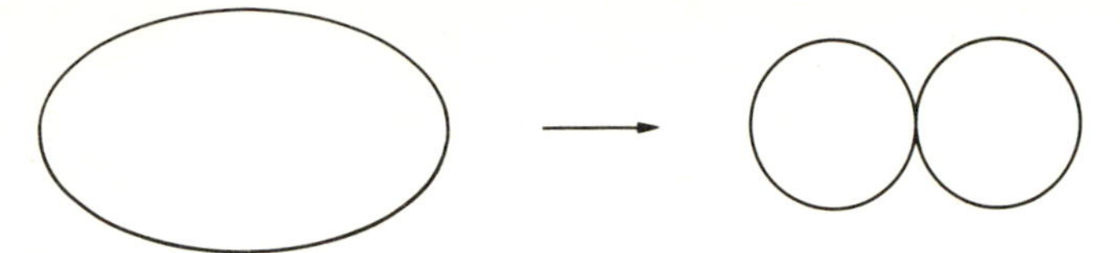

FIG. 3.11. Symmetric fission to two touching spherical daughter nuclei.

energy has only increased by ~ 25 per cent, while the Coulomb energy has fallen to only ~ 16 per cent of its value for the parent sphere. Hence the energy of the drop during the fissioning process must have the form illustrated in Fig. 3.12. The exact structure of the fission barrier we shall discuss in more detail in Chapter 7.

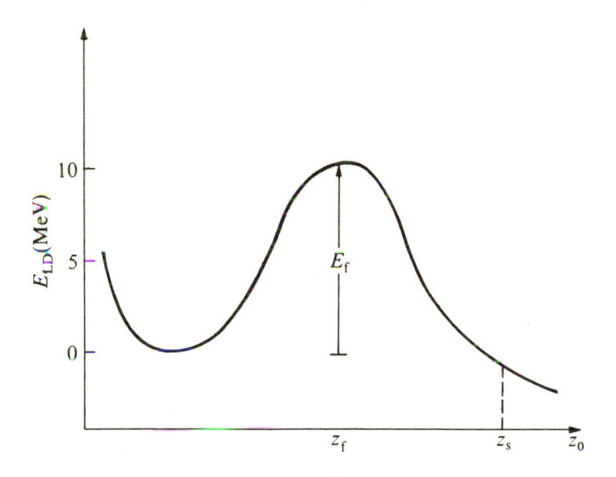

FIG. 3.12. The fission barrier.

Calculations of liquid-drop energies and equilibrium shapes have been carried out for generalizations of eqn (3.41)

$$R = R_0\left(1 + \sum_{n=2}^{N} \alpha_n P_n(\cos\theta)\right), \tag{3.43}$$

with N as high as 18. The fissioning process has been described in terms of the quantum-mechanical tunnelling of the fission fragments through the fission barrier in the N- dimensional generalized coordinate space.

Geometrically, it appears more natural to discuss the shapes approached just before fission by a two-centre model, e.g. the intersecting spheroids of Fig. 3.13 in which the critical variational parameter is the separation between the centres and the remaining parameters are optimized at each separation distance subject to the conditions of nuclear incompressibility. The details of such calculations will be discussed in Chapter 7.

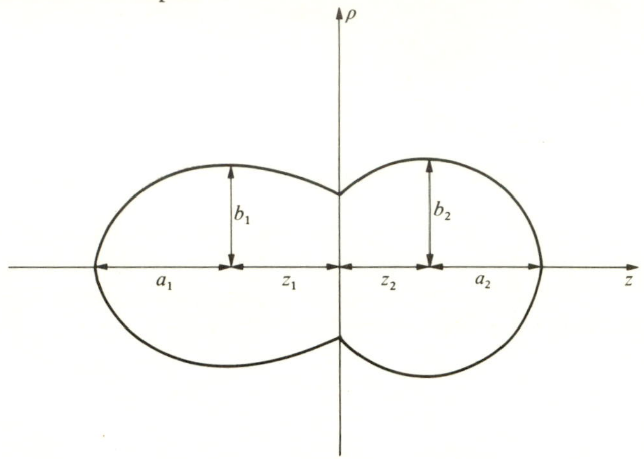

FIG. 3.13. A two-centre nuclear model.

It should be clear, however, that nothing as simplified as the extreme liquid-drop model can hope to account for the difference in the odd-isotope and even-isotope fission characteristics. Nor, as we shall see, can it account for the observed asymmetric fission yields.

3.3 The individual-particle model

3.3.1. Origins of the shell model (Brown 1967; Irvine 1972)

If we now take the microscopic view that the nucleus is composed of N neutrons and Z protons interacting with each other in a pairwise fashion through a two-body potential V_{ij} which may be deduced from the nucleon–nucleon scattering data, then the system will be described by the two-body Hamiltonian

$$\mathcal{H} = \sum_{i=1}^{A} -\frac{\hbar^2}{2m}\nabla_i^2 + \sum_{i>j}^{A} V_{ij}. \tag{3.44}$$

As is usual in many-body systems our first approximation is to consider the system as composed of independent particles each moving in the mean field U_i provided by its neighbours,

$$\mathcal{H} = \sum_{i=1}^{A}\left(-\frac{\hbar^2}{2m}\nabla_i^2 + U_i\right) + \sum_{i>j}^{A} V_{ij} - \sum_{i=1}^{A} U_i = \mathcal{H}_0 + \mathcal{H}_1. \tag{3.45}$$

$\mathcal{H}_1$ we will refer to as the residual interaction and will be treated as a perturbation on the eigenstates of $\mathcal{H}_0$.

Since $\mathcal{H}_0$ is an effective one-body operator its eigenstates are simply products of one-body wavefunctions. Because the nucleons are fermions these product wavefunctions must be antisymmetrized, i.e. Slater determinants

$$\mathcal{H}_0 \, \Phi_n = E_n \, \Phi_n, \tag{3.46}$$

where

$$\Phi_n = \frac{1}{\sqrt{A!}} \det \phi_i(x_j) \tag{3.47}$$

and

$$E_n = \sum_{i \leqslant \epsilon_n} \epsilon_i, \tag{3.48}$$

with

$$\left(-\frac{\hbar^2}{2m} \nabla^2 + U \right) \phi_i = \epsilon_i \, \phi_i. \tag{3.49}$$

If U is to represent the field produced on a given nucleon by its neighbours, then we have

$$U_i = \sum_{j \neq i} V_{ij}. \tag{3.50}$$

However, as so defined, U_i is still intrinsically a two-body operator, and in order to obtain the simple local effective one-body operator suitable for eqn (3.49) we must average U_i over the nucleons in the nucleus

$$U(x) = \frac{\sum_{ij} \int \phi_i^\star(x) \, \phi_j^\star(x') \, V(xx') \, (\phi_i(x)\phi_j(x') - \phi_i(x')\phi_j(x)) \, \mathrm{d}x'}{\sum_i |\phi_i(x)|^2}. \tag{3.51}$$

We are now clearly faced with a self-consistency problem, i.e. the single-particle potential U depends on the wavefunctions ϕ (see eqn (3.51)), which are in turn determined by the single-particle potential U (see eqn (3.49)). We shall not consider in detail the solution to this problem, but we shall restrict ourselves to a few general observations about the nature of the mean field:

1. If the number of neutrons equals the number of protons and they occupy the same single-particle states then the mean field will be the same for neutrons and protons (neglecting the Coulomb force), thus an isotopic-spin formalism will be valid. None of these conditions are met in the heavy-mass region.

2. If the mean field is spherically symmetric then the one-body eigenstates of eqn (3.49) will be eigenstates of angular momentum and we shall have a shell structure. However, only a closed-shell configuration is spherical. Hence for a closed-shell configuration the assumption of a spherical mean field will be self-consistent. For other configurations involving partially filled shells it will not in general be self-consistent.
3. Point 2 tells us something about the angular variation of the mean field. However the details of this and the radial form are dependent on the two-body interaction V. We note that this is short-ranged, and if it can be approximated by a δ-function form with an exchange mixture, i.e.

$$V(\mathbf{x}_1, \mathbf{x}_2) = -V_0(W + MP_X + BP_\sigma + HP_\tau)\,\delta(r_{12}), \tag{3.52}$$

then eqn (3.51) reduces to

$$U_C = -U_0 \sum_i |\phi_i(x)|^2 = -U_0\rho(x), \tag{3.53}$$

where ρ is the density of the nucleus and the constant U_0 depends on the exchange mixture parameters W, M, B, H and the strength of the two-nucleon interaction V_0. Thus in the limit of a short-range interaction the self-consistent mean field is proportional to the density. Since the density is experimentally measurable we can phenomenologically obtain the functional form of the mean field.

Point 2 is consistent with point 3 but is not restricted to the short-range limit of the nucleon–nucleon interaction.

4. The two-body interaction is spin-dependent, and besides the exchange mixture of eqn (3.52) it contains a spin–orbit interaction,

$$V_{LS} = V_0{}^{LS}\,\mathbf{L}\,.\,\mathbf{S}\,. \tag{3.54}$$

If we now require about the state of motion of a single particle outside a closed-shell configuration, we have that the mean field due to the central interaction is spherically symmetric, and inserting (3.54) into (3.50) we see that the summation is over closed-shell configurations. Hence there is a complete cancellation between the spins and angular momenta of the core states, resulting in a non-central mean field of the form

$$U_{LS} = \xi\,(\mathbf{r}_i)\,\mathbf{l}_i.\,\mathbf{s}_i, \tag{3.55}$$

i.e. a one-body spin–orbit force acting on the extra core particle. Allowing for the possibility of a tensor force and a quadratic spin–orbit force we can obtain an additional non-central force of the form

$$U_{L^2} = \eta(\mathbf{r}_i)\,\mathbf{l}_i^2\,. \tag{3.56}$$

Thus the mean field seen by a nucleon due to closed shells is spherically symmetric and of the form

$$U(r) = -U_0\rho(r) + \xi(r)\,\mathbf{l}\,.\,\mathbf{s} + \eta(r)\,\mathbf{l}^2\,, \tag{3.57}$$

where $\rho(r)$ is the density of the closed-shell configuration. Given U then the residual interaction is defined by eqn (3.45). The form (3.57) is adequate for a few particles outside a closed-shell configuration. The contribution of the valence particles to their mutual mean field may be small and treated as a perturbation, i.e. via the residual interaction. However, when there are a large number of valence particles their contribution can become sizeable and must be added to $U(r)$. As we have already noted the contribution of such open-shell configurations is in general non-spherical. Clearly, depending on our choice of U, we obtain different residual interactions (see eqn (3.45)).

There is one more general point we must consider before we proceed to apply our microscopic model to the heavy nuclei. We know from the two-nucleon scattering data that the neucleon–nucleon interaction has a strong repulsive core at short distances (Fig. 3.14), and if this goes to infinity faster than $1/r$,

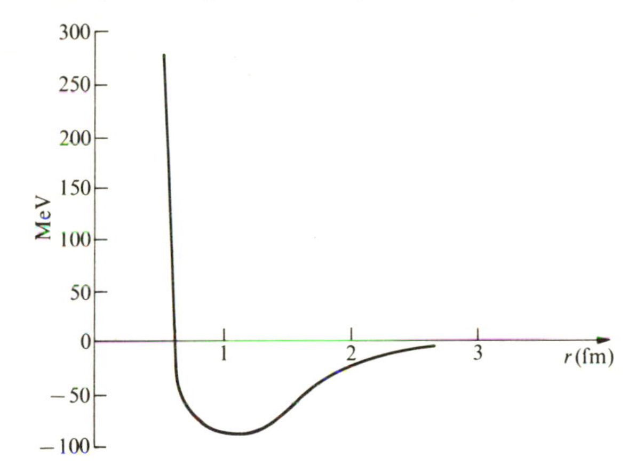

FIG. 3.14. Schematic view of the nucleon–nucleon potential.

then the integral in (3.51) becomes infinite. This is simply a mathematical consequence of the physical situation in which our basis Slater determinantal wavefunctions (3.47) are truly independent-particle wavefunctions in which the nucleons move freely throughout the nuclear volume constrained only by the mean field of their interaction with the other nucleons. However, if the nucleons have a hard core then there are regions where a given nucleon cannot go, i.e. where there is already another nucleon. Such an effect is a two-body correlation and is not contained in the determinantal wavefunctions. The simplest modification which will accommodate these correlations is

$$\Psi_n \rightarrow F \det \phi_i(\mathbf{x}_j), \tag{3.58}$$

where

$$F = \prod_{i>j} f(r_{ij}) \tag{3.59}$$

contains the two-body correlation function

$$f(r) = 1 - g(r) = 0, \; r < r_{\text{core}}. \\ \rightarrow 1, \; r \rightarrow \infty \tag{3.60}$$

If the nuclear wavefunction heals quickly, i.e. $f(r) \rightarrow 1$ rapidly for $r > r_{\text{core}}$, then $g(r)$ will be an extremely short-ranged function differing from zero only for $r \lesssim r_{\text{core}}$. If the nuclear density is low in the sense that the excluded volume is only a small fraction of the total nuclear volume then we may approximate F by

$$F \rightarrow 1 - \sum_{i>j} g(r_{ij}). \tag{3.61}$$

Then, assuming that the density is low and ignoring clusters of more than two particles, we find that

$$\left\langle \Psi_n \,\middle|\, \sum_{i>j} V_{ij} \,\middle|\, \Psi_n \right\rangle = \left\langle \Phi_n \,\middle|\, F \sum_{i>j} V_{ij} F \,\middle|\, \Phi_n \right\rangle \rightarrow \left\langle \Phi_n \,\middle|\, \sum_{i>j} f(r_{ij})\, V(r_{ij})\, f(r_{ij}) \,\middle|\, \Phi_n \right\rangle. \tag{3.62}$$

Thus we have an effective nucleon–nucleon interaction of the form

$$V_{\text{eff}} = f\,V\,f \tag{3.63}$$

in which the correlation functions suppress the repulsive core of the bare nucleon–nucleon interaction. We can then proceed as outlined above using this effective interaction. There is a formal perturbation theory approach to strongly interacting systems called Brueckner–Bethe–Goldstone theory, which achieves essentially the same result but which we shall not discuss in detail here. The low-density approximation is valid provided

$$\kappa = (r_{\text{core}}/r_0)^3 \ll 1; \tag{3.64}$$

with $r_{\text{core}} \simeq 0{\cdot}4$ fm and $r_0 \simeq 1{\cdot}2$ fm we have $\kappa \simeq 1/27$ and hence our approximation should be valid to about 10 per cent.

Having justified the microscopic model we shall now adopt a largely phenomenological approach in which the mean field U will be approximated by a form chosen either to produce analytic eigenstates which are easy to manipulate or to produce eigenstates which reproduce some observed features of nuclear structure which are under study. In this situation the residual interaction is extremely uncertain, and it is usually replaced by a phenomenological two-body interaction bearing little or no resemblance to that deduced from the nucleon–nucleon scattering data.

3.3.2 The spherical shell model (De Shalit and Talmi 1963)

As we have seen for a near-closed-shell nucleus the mean field is spherically symmetric and of the form

$$U(r) = -U_0\rho(r) + \xi(r)\,\mathbf{l}\,.\,\mathbf{s} + \eta(r)\,\mathbf{l}^2. \tag{3.65}$$

This leads to eigenstates labelled by the total single-particle angular momentum $\mathbf{j} = \mathbf{l} + \mathbf{s}$ (where s is spin-$\frac{1}{2}$ for nucleons), the component m_j of $\mathbf{j}$ along an arbitrary *z*-axis, the magnitude of the orbital angular momentum l and a label denoting neutron or proton τ, plus a principal quantum number n denoting the number of nodes in the single-particle wavefunction. The single-particle energies $\epsilon_{njl\tau}$ are functions only of n, j, l and τ and are $(2j + 1)$ degenerate in the quantum number m_j. Fig. 3.15 ilustrates the form of the spherical shell model potential and the

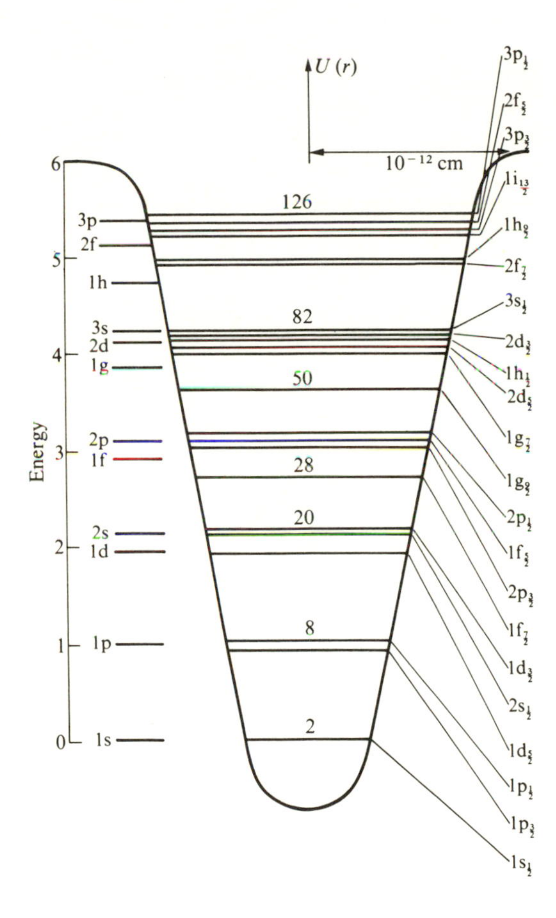

FIG. 3.15. Single-particle energies in a typical nuclear mean field. At the extreme left the major shells are represented without spin–orbit splitting. The energy scale is arbitrary.

resulting single-particle energies. Configurations corresponding to the closure of j shells reproduce the magic numbers referred to in section 3.2.1. In particular we see that the ground state of ^{208}Pb is predicted to be a doubly magic nucleus corresponding to the closed neutron shell at $N = 126$ and the closed proton shell with $Z = 82$. The addition of a single neutron to form ^{209}Pb or a single proton to form ^{209}Bi should not greatly affect the mean field, and hence the extra particle would be expected to occupy the single-particle states of the spherically symmetric field U outside an inert doubly closed-shell ^{208}Pb core. In this approximation the excitation energies of the mass-209 isobars are given by

$$E_i(209) = E(^{208}\mathrm{Pb}) + \epsilon_{njl_\tau}. \tag{3.66}$$

Fig. 3.16 illustrates how this interpretation may be used to identify the single-particle eigenstates in the mean field of ^{208}Pb. The separation energies of the last

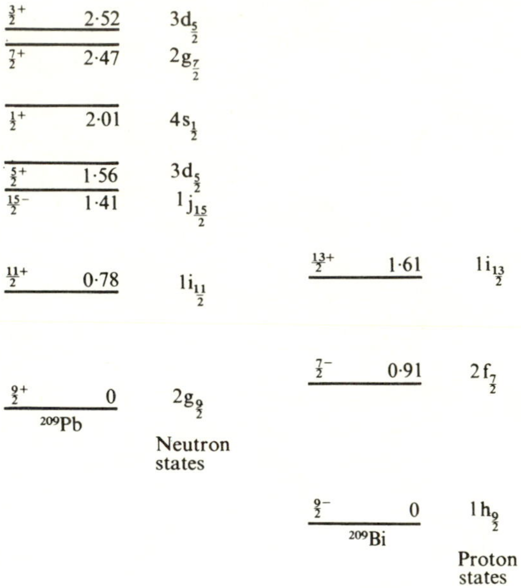

J^π	Energy	State
$\frac{3}{2}^+$	2·52	$3d_{\frac{3}{2}}$
$\frac{7}{2}^+$	2·47	$2g_{\frac{7}{2}}$
$\frac{1}{2}^+$	2·01	$4s_{\frac{1}{2}}$
$\frac{5}{2}^+$	1·56	$3d_{\frac{5}{2}}$
$\frac{15}{2}^-$	1·41	$1j_{\frac{15}{2}}$
$\frac{11}{2}^+$	0·78	$1i_{\frac{11}{2}}$
$\frac{9}{2}^+$	0	$2g_{\frac{9}{2}}$

^{209}Pb — Neutron states

J^π	Energy	State
$\frac{13}{2}^+$	1·61	$1i_{\frac{13}{2}}$
$\frac{7}{2}^-$	0·91	$2f_{\frac{7}{2}}$
$\frac{9}{2}^-$	0	$1h_{\frac{9}{2}}$

^{209}Bi — Proton states

FIG. 3.16. Shell-model identification of low-lying single-particle states in ^{209}Pb and ^{209}Bi.

nucleon in $A = 209$ nuclei gives information about U_0 and $\eta(r)$, while information about the spin–orbit splitting function $\xi(r)$ can be obtained from the excitation spectra.

Typically we have

$$U = -U_0 f(r, r_0, a) + U_{\mathrm{SO}} \frac{1}{r}\frac{\partial f}{\partial r} \mathbf{l} \cdot \mathbf{s} + V_{\mathrm{C}}, \tag{3.67}$$

with

$$f(r, r_0, a) = [1 + \exp\{-(r - r_0 A^{\frac{1}{3}})/a\}]^{-1}, \tag{3.68}$$

where Greenlees *et al* (1970) give U_0 = 52 MeV, U_{so} = 5·7 MeV, r_0 = 1·15 fm, and a = 0·6 fm, and V_C is the Coulomb potential due to a charge Ze distributed uniformly throughout a volume of the form $f(r, r_0, a)$.

For nuclei more massive than A = 209 we assume that the (A − 208) valence particles are distributed over the single-particle levels of Fig. 3.16. Each distribution giving rise to a determinantal basis function Φ_n. The unperturbed energy of each such configuration will be

$$E_0 = E(^{208}\text{Pb}) + \sum_{i=209}^{A-208} \epsilon_i. \tag{3.69}$$

Because of the $(2j + 1)$ degeneracy of a single-particle level of angular momentum j and the relatively high density of single-particle states evident in Fig. 3.16, there will be a large number of degenerate or near-degenerate configurations Φ_n. We now form trial variational wavefunctions Ψ_n as linear combinations of these basis states,

$$\Psi_n = \sum_m c_{nm}\, \Phi_m, \tag{3.70}$$

and within this basis we obtain the eigenenergies and eigenstates by diagonalizing the matrix $\langle \Phi_n \mid \mathcal{H} \mid \Phi_m \rangle$ using the residual interaction to mix configurations. In constructing this matrix we must severely limit the sum in eqn (3.70) to include a small subset of configurations which we think will be most important. A restriction to a subset of single-particle energy levels in the valence major shell is the most usual assumption. Even with this restriction the size of the matrix rapidly becomes enormous for anything beyond A = 210 or 211. However, we may make use of the Racah angular momentum algebra to form a new basis set of states coupled to good angular momentum and parity

$$\Phi'_m(JM) = \sum_n C^J_{nm}\, \Phi_n, \tag{3.71}$$

and in this basis the matrix $\langle \Phi'_m \mid \mathcal{H} \mid \Phi'_n \rangle$ breaks into block diagonal form and the individual blocks can be diagonalized separately (see Fig. 3.17). However, in the heavy nuclei there are ~ 58 neutron orbits in the major neutron valence shell and ~ 44 proton orbits in the major proton valence shell (see Fig. 3.15), and in a typical heavy element, say ^{236}U, there are 10 valence protons and 18 valence neutrons. The resulting number of configurations is so large that conventional shell-model calculations for such heavy nuclei are not at all practicable.

Indeed it is not sensible to carry out such calculations in a spherical basis. The approximation of a spherical mean field will be valid only if the number of valence particles is $\lesssim$ 10 per cent of the number of core particles producing a spherical field. With a ^{208}Pb core we might expect that the spherical basis would cease to be appropriate for A between 220 and 230. Note this is exactly where there is evidence that the deformed rotational spectrum is taking over from the spherical vibrational spectrum.

3.3.3. The deformed shell model (*The Nilsson scheme*) (Davidson 1968; Nilsson 1955)

We saw in section 3.3.1 that in the limit of an extremely short-range interaction the self-consistent mean field has the same form as the density. Further, Fig. 2.5 and section 3.2.3 have indicated that the observed spectra of heavy nuclei $A \gtrsim 230$ are consistent with a spheroidal density distribution. In such a potential the angular momentum of the individual particles is no longer a conserved quantity. However, such a potential has an axis of symmetry, and the component k of angular

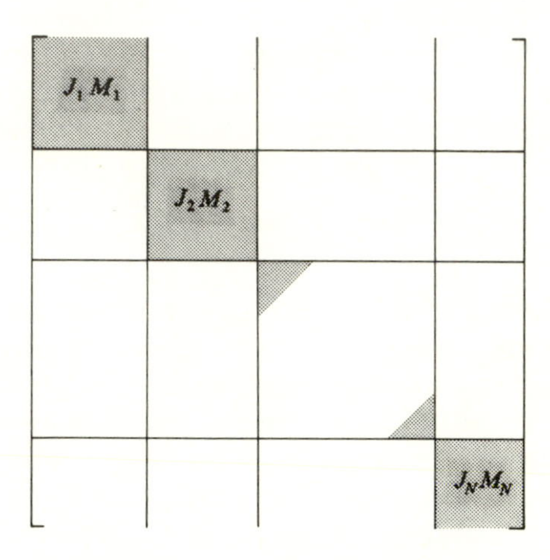

FIG. 3.17. The nuclear Hamiltonian matrix breaks into block diagonal form in the basis of states of good angular momentum, etc.

momentum along that axis will be a good quantum number. Also a spheroidal potential enjoys reflection symmetry and hence the parity p will be conserved. There is also a symmetry plane normal to the axis of symmetry and hence the single-particle levels with components k and $-k$ of angular momentum will be degenerate. Thus the basis states, i.e. Slater determinants of single-particle states in the spheroidal well, will be labelled by the total component of angular momentum K along the symmetry axis,

$$K = \sum_{i=1}^{A} k_i, \tag{3.72}$$

and the parity P,

$$P = \prod_{i=1}^{A} p_i. \tag{3.73}$$

Because the levels are pairwise degenerate the unperturbed ground state of any even–even nucleus will have

$$K = 0, \quad P = +. \tag{3.74}$$

For an odd-A nucleus we shall have

$$K = k(\text{last nucleon}), \quad P = p(\text{last nucleon}), \tag{3.75}$$

and for an odd–odd nucleus

$$K = k(\text{last proton}) + k(\text{last neutron})$$

$$P = p(\text{last proton}) \times p(\text{last neutron}). \tag{3.76}$$

In order to make any additional statements about nuclear structure we require a specific form for the mean field. Thus we shall have a single-particle Hamiltonian $\mathcal{H}_0(\{\alpha\}, \{\beta\})$ which will be a function of two sets of parameters $\{\alpha\}$ and $\{\beta\}$. The parameters $\{\alpha\}$ will be fitted to some experimental information while the $\{\beta\}$ will be treated variationally. The Nilsson scheme provides us with the simplest example of such an approach.
Here we assume

$$\mathcal{H}_0 = \sum_{i=1}^{A} \{h_c\ (i) + h_{nc}\ (i) + h_\delta\ (i)\}, \tag{3.77}$$

where

$$h_c(i) = \tfrac{1}{2}\hbar\,\omega_0(-\nabla_i^2 + r_i^2), \tag{3.78}$$

$$h_{nc}\ (i) = \mathrm{C}\ \mathbf{l}_i\ .\ \mathbf{s}_i + \mathrm{D}\mathbf{l}_i^2, \tag{3.79}$$

and

$$h\delta = -\,\delta\hbar\,\omega_0\ \tfrac{4}{3}\sqrt{\left(\frac{\pi}{4}\right)}\,r_i^2\ Y_{20}\ (\Omega_i) \tag{3.80}$$

This should be compared with eqn (3.65).
Throughout the position parameters r_i are dimensionless quantities measured in units of the oscillator size parameter $b = \sqrt{(\hbar/m\omega_0)}$. The set of parameters $\{\alpha\}$ now comprise b†, C and D, where b is adjusted to give agreement between the model and the observed radius of nuclear charge distributions, while C and D are adjusted to reproduce the single-particle spacings observed near closed-shell nuclei (see Fig. 3.16). The set of parameters β comprises the single element δ, which

† For nuclear imcompressibility we require that ω_0 be a function of δ with

$$\omega_0(\delta) = \omega_0(0)\ (1 - \tfrac{4}{3}\delta^2 - \tfrac{16}{27}\delta^3)^{-\frac{1}{6}}.$$

Matching the model nuclear radius to observed charge distributions yields

$$\hbar\omega_0\ (0) \simeq 41\mathrm{A}^{-\frac{1}{3}}.$$

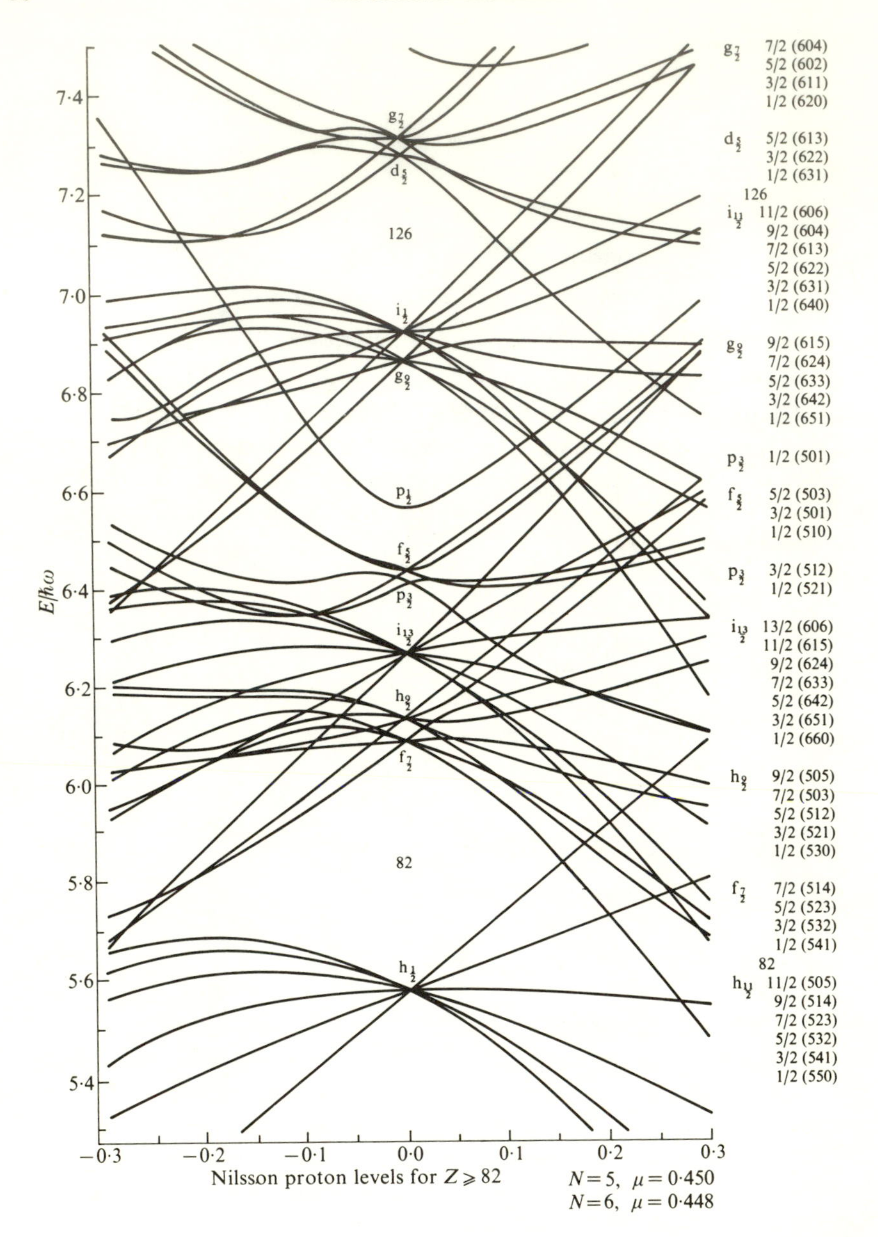

FIG. 3.18. (a)

measures the quadrupole deformation of the oscillator well. This can be related to the deformation parameter β of the liquid-drop model (eqn (3.34))

$$\delta \simeq \tfrac{3}{2}\sqrt{\left(\frac{5}{4\pi}\right)}\,\beta \simeq 0{\cdot}95\,\beta. \tag{3.81}$$

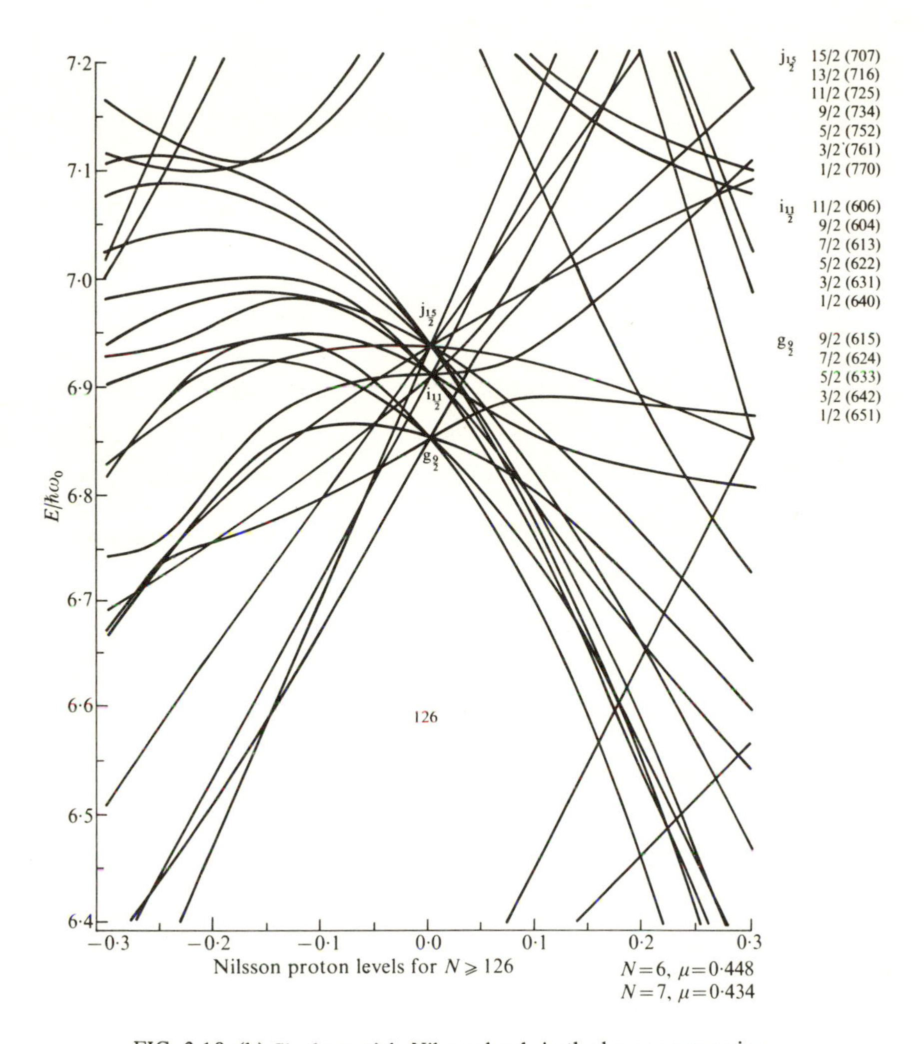

FIG. 3.18. (b) Single-particle Nilsson levels in the heavy-mass region.

As discussed above, the single-particle energies can be classified by four quantum numbers n, k, p, and τ – i.e. an energy-ordering principal quantum number, the component of angular momentum along the axis of symmetry, the parity of the state, and whether it is a neutron or proton state. The states $\phi_{nkp\tau}$ and $\phi_{n-kp\tau}$ will be degenerate, and the energies $\epsilon_{nkp\tau}$ will be functions of the nuclear deformation δ.

The expectation value of the Hamiltonian (3.44) in a Slater determinant configuration is

$$E_0 = \sum_i h\omega_0 \langle i \,|\nabla^2\,| i\rangle + \sum_{i>j} \langle ij \,|\, V\,[\,1_{ij}\rangle - 1_{ji}\rangle], \qquad (3.82)$$

where the summations are over the states occupied in the determinant. Assuming (3.49) and (3.50), eqn (3.82) may be rewritten

$$E_0\,(\delta) = \tfrac{1}{2} \sum_i \epsilon_{n_i k_i p_i \tau_i}\,(\delta) - \tfrac{1}{2} \sum_i \langle i \,|\nabla^2\,| i\rangle \hbar\omega_0\,(\delta)$$

$$= \tfrac{1}{2} \sum_i \epsilon_i\,(\delta) - \tfrac{1}{4} \sum_i (2n_i + l_i + \tfrac{3}{2})\,\hbar\omega_0\,(\delta), \qquad (3.83)$$

and the equilibrium deformation can be obtained from the variational principle

$$\partial E_0\,(\delta)/\partial\delta = 0. \qquad (3.84)$$

In Fig. 3.18 we plot the numerically calculated single-particle energies ϵ_i as a function of the deformation parameters for the heavy-mass region. The corresponding single-particle eigenstates can be expressed as a linear combination of the $\delta = 0$ spherical-oscillator basis states,

$$\phi_{nkp\tau} = \sum_{n'jl} C^{nkp\tau}_{n'jl}\, \phi^{\mathrm{HO}}_{n'(\frac{1}{2}\,l)jk\tau}\,, \quad p = (-1)^l. \qquad (3.85)$$

The most common approximation is to restrict the summation in eqn (3.85) to a single major oscillator shell. Since in each major shell the value of j uniquely determines the value of ℓ and we assume that parity and charge are conserved, we may write simply

$$\phi_{nk} = \sum_j C^{nk}_j\, \phi_{njk}. \qquad (3.86)$$

In Table 3.4 are shown the numerically determined coefficients C^{nk}_j and corresponding eigenenergies ϵ_{njk} for the heavy nuclei.

In discussing deformations we have so far restricted ourselves to simple quadrupole distortions of an oscillator well. Clearly, however, one could generalize this, as has been done in the case of the liquid-drop model (see eqns (3.26) and (3.43)), solving numerically for the single-particle eigenstates and eigenenergies in the distorted well and determining the distortion parameters α_n variationally. In a study of fission it may be more appropriate to define a two-centred single-particle potential well. We shall discuss this generalization in Chapter 7.

Whatever the choice of single-particle wavefunctions the resulting basis determinantal wavefunctions Φ_{KP} do not have good angular momentum. At most they will be labelled by the total component of angular momentum K along the symmetry axis and the parity P. To obtain a comparison with experiment we must project out states of good angular momentum (MacDonald 1970)

$$\Phi^{K}_{JMP} = \int \mathcal{D}^{J}_{MK}(\alpha, \beta, \gamma)\, \Phi_{KP}(\alpha, \beta, \gamma)\, \mathrm{d}\alpha\, \mathrm{d}\beta\, \mathrm{d}\gamma, \tag{3.87}$$

where α, β, and γ are the Euler angles of a body-fixed axes relative to a space-fixed frame. Strictly speaking, this projection should be carried out before the deformation parameters are determined variationally. However, if the nuclear shape does not depend strongly on J, e.g. as in the rotations of the rigid liquid-drop shape of section 3.2.3, then the optimum deformation parameters will be the same whether the projection is carried out before or after the variational calculation. That the projected states Ψ^{K}_{JMP} can have a spectrum similar to the rotational bands of section 3.2.3 we shall demonstrate in Chapter 5.

TABLE 3.4

Nilsson single-particle energies and eigenstates†

N = 5 KAPPA = 0.050 0, MU = 0.4500

DELTA =		−0.3	−0.2	−0.1	0.0	0.1	0.2	0.3
K = 11/2, (505)								
E		5.075 000	5.241 667	5.408 333	5.575 000	5.741 667	5.908 333	6.075 000
L	J							
5	11/2	1.000 000	1.000 000	1.000 000	1.000 000	1.000 000	1.000 000	1.000 000
K = 9/2, (514)								
E		5.325 000	5.415 110	5.497 484	5.575 000	5.649 452	5.721 913	5.793 029
L	J							
5	9/2	.254 824	.144 135	.061 060	.000 000	−0.045 360	−0.079 796	−0.106 565
5	11/2	.966 988	.989 558	.998 134	1.000 000	.998 971	.996 811	.994 306
K = 9/2, (505)								
E		5.675 000	5.818 223	5.969 183	6.125 000	6.283 881	6.444 754	6.606 971
L	J							
5	9/2	.966 988	.989 558	.998 134	1.000 000	.998 971	.996 811	.994 306
5	11/2	−0.254 824	−0.144 135	−0.061 060	.000 000	.045 360	.079 796	.106 565
K = 7/2, (523)								
E		5.430 150	5.518 608	5.561 996	5.575 000	5.571 353	5.558 289	5.539 509
L	J							
3	7/2	.608 688	.372 199	.146 739	.000 000	−0.088 767	−0.144 674	−0.181 890
5	9/2	.041 626	.070 955	.050 350	.000 000	−0.053 147	−0.099 469	−0.137 603
5	11/2	.792 317	.925 437	.987 893	1.000 000	.994 634	.984 467	.973 643
K = 7/2, (514)								
E		5.723 873	5.813 851	5.937 214	6.080 000	6.169 566	6.224 412	6.281 708

†From Irvine (1972, Table 16.1).

TABLE 3.4 (*contd.*)

N = 5 KAPPA = 0.0500 MU = 0.4500

L	J							
3	7/2	.724 457	.885 300	.966 306	1.000 000	.367 229	.250 514	.205 741
5	9/2	−0.436 348	−0.326 610	−0.220 804	.000 000	.926 484	.958 837	.962 919
6	11/2	−0.533 631	−0.331 014	−0.132 279	.000 000	.082 280	.133 695	.174 522

K = 7/2, (503)

	E	6.025 977	6.047 541	6.080 790	6.125 000	6.239 080	6.397 300	6.558 783
L	J							
3	7/2	.323 513	.278 769	.211 471	.000 000	.925 885	.957 242	.961 554
5	9/2	.898 815	.942 492	.974 018	1.000 000	−0.372 562	−0.265 965	−0.232 062
5	11/2	−0.295 756	−0.184 380	−0.081 054	.000 000	.062 724	.113 800	.146 835

K = 5/2, (532)

	E	5.558 017	5.603 131	5.608 830	5.575 000	5.509 241	5.421 647	5.320 040
L	J							
3	5/2	.196 137	.081 353	.017 424	.000 000	.010 365	.030 947	.052 573
3	7/2	.517 285	.378 086	.188 928	.000 000	−0.148 029	−0.250 604	−0.319 443
5	9/2	.096 425	.065 238	.038 123	.000 000	−0.049 015	−0.099 174	−0.143 829
5	11/2	.827 435	.919 879	.981 096	1.000 000	.987 713	.962 499	.935 150

K = 5/2, (523)

	E	5.920 678	6.035 791	6.062 529	6.080 000	6.069 661	6.024 968	5.980 368
L	J							
3	5/2	.814 929	.719 005	.136 848	.000 000	−0.171 350	−0.240 247	−0.272 450
3	7/2	.081 919	.469 871	.950 526	1.000 000	.571 795	.365 383	.276 246
5	9/2	.488 269	.424 263	−0.215 428	.000 000	.792 219	.878 288	.888 147
5	11/2	−0.301 286	−0.286 802	−0.177 101	.000 000	.126 807	.193 356	.246 281

K = 5/2, (512)

	E	6.087 094	6.071 566	6.127 918	6.125 000	6.129 193	6.175 570	6.230 587
L	J							
3	5/2	.104 027	.164 249	.386 994	.000 000	−0.000 108	−0.062 070	−0.100 600
3	7/2	.720 664	−0.702 587	.137 543	.000 000	.800 573	.882 940	.886 379
5	9/2	−0.548 322	.653 808	.909 175	1.000 000	−0.592 348	−0.425 690	−0.376 047
5	11/2	−0.411 296	.227 881	−0.068 687	.000 000	.090 588	.188 023	.250 601

K = 5/2, (503)

	E	6.444 211	6.366 179	6.344 055	6.430 000	6.568 572	6.721 148	6.879 005
L	J							
3	5/2	−0.535 350	.670 400	.911 704	1.000 000	.985 156	.968 231	.955 458
3	7/2	.454 256	−0.377 683	−0.204 669	.000 000	.101 098	.155 274	.189 677
5	9/2	.672 041	−0.623 122	−0.354 313	.000 000	.138 243	.193 810	.221 578
5	11/2	−0.235 401	.140 137	0.36 989	.000 000	.011 674	.029 267	.045 158

K = 3/2, (541)

	E	5.613 658	5.653 436	5.639 217	5.575 000	5.465 398	5.318 262	5.143 708
L	J							
1	3/2	.332 471	.147 684	.033 071	.000 000	.021 480	.064 311	.106 926
3	5/2	−0.001 378	.021 449	.010 296	.000 000	.013 418	.048 487	.091 702

TABLE 3.4 (*contd.*)

N = 5 KAPPA = 0.0500 MU = 0.4500

L	J							
3	7/2	.542 245	.407 378	.213 661	.000 000	−0.194 831	−0.343 046	−0.442 109
5	9/2	.011 602	.030 399	.023 022	.000 000	−0.035 139	−0.079 198	−0.124 203
5	11/2	.771 555	.900 471	.976 022	1.000 000	.979 880	.932 502	.877 081
K = 3/2, (532)								
E		5.938 348	6.035 062	6.106 163	6.080 000	5.988 483	5.850 992	5.706 138
L	J							
1	3/2	.698 921	.707 162	.447 905	.000 000	−0.190 101	−0.204 870	−0.195 516
3	5/2	−0.404 870	−0.288 552	−0.064 405	.000 000	−0.240 178	−0.395 918	−0.460 196
3	7/2	.187 988	.458 670	.835 777	1.000 000	.730 360	.436 783	.285 319
5	9/2	−0.358 518	−0.336 302	−0.244 883	.000 000	.585 313	.737 507	.751 517
5	11/2	−0.428 621	−0.305 258	−0.191 681	.000 000	.173 665	.258 034	.322 194
K = 3/2, (521)								
E		6.182 146	6.178 219	6.171 720	6.125 000	6.041 810	5.981 257	5.932 175
L	J							
1	3/2	.339 115	.213 351	.027 764	.000 000	.126 737	−0.257 967	−0.323 225
3	5/2	.499 172	.482 689	.345 800	.000 000	−0.123 148	.050 140	−0.001 613
3	7/2	.276 977	.357 867	.256 248	.000 000	−0.603 738	.742 026	.727 214
5	9/2	.660 855	.734 202	.898 490	1.000 000	.771 711	−0.568 139	−0.504 654
5	11/2	−0.349 832	−0.233 175	−0.081 877	.000 000	−0.093 460	.239 906	.334 675
K = 3/2, (512)								
E		6.490 465	6.386 289	6.336 186	6.405 000	6.406 728	6.433 825	6.478 118
L	J							
1	3/2	−0.380 012	−0.531 235	.803 539	1.000 000	.428 090	.338 088	.282 135
3	5/2	.365 975	.359 176	−0.389 278	.000 000	.850 503	.824 538	.800 601
3	7/2	.635 077	.605 968	−0.378 515	.000 000	.202 534	.296 174	.344 242
5	9/2	−0.492 924	−0.434 579	.237 875	.000 000	.227 184	.336 196	.385 778
5	11/2	−0.274 512	−0.180 902	.054 129	.000 000	.027 387	.071 319	.110 050
K = 3/2, (501)								
E		6.890 383	6.695 328	6.528 381	6.430 000	6.545 914	6.697 331	6.854 861
L	J							
1	3/2	.376 251	.387 791	.389 669	.000 000	.874 115	.879 228	.875 366
3	5/2	.673 029	.744 500	.851 253	1.000 000	−0.451 235	−0.398 147	−0.372 620
3	7/2	−0.436 568	−0.358 328	−0.216 539	.000 000	.151 971	.230 691	.275 300
5	9/2	−0.437 752	−0.397 574	−0.275 015	.000 000	−0.094 994	−0.118 330	−0.127 654
5	11/2	.152 472	.094 197	.031 706	.000 000	0.13 828	.034 882	.052 935
K = 1/2, (550)								
E		5.653 002	5.679 849	5.654 246	5.575 000	5.442 396	5.258 287	5.028 206
A		−4.918 695	−5.550 135	−5.891 695	−6.000 000	−5.893 002	−5.568 550	−5.043 631
L	J							
1	1/2	.143 797	.047 587	.006 086	.000 000	−0.005 072	−0.032 821	−0.082 498
1	3/2	.278 157	.145 646	.039 753	.000 000	.037 838	.131 479	.239 157
3	5/2	.065 981	.021 448	.004 400	.000 000	.007 260	.037 276	.091 462
3	7/2	.536 593	.413 842	.224 575	.000 000	−0.222 178	−0.405 078	−0.524 934
5	9/2	.033 073	.015 051	.007 875	.000 000	−0.013 110	−0.035 777	−0.069 768
5	11/2	.780 110	.896 979	.973 585	1.000 000	.974 143	.902 706	.804 493

TABLE 3.4 (*contd.*)

N = 5 KAPPA = 0.0500 MU = 0.4500

K = 1/2, (541)

L	J							
E		6.006 634	6.107 475	6.143 242	6.080 000	5.936 562	5.713 421	5.462 905
A		.774 817	−0.117 957	−3.646 786	−4.000 000	−2.707 605	.309 957	1.043 088
1	1/2	.601 616	.498 808	.138 630	.000 000	.113 594	.268 673	−0.329 596
1	3/2	.338 085	.453 905	.409 340	.000 000	−0.365 483	−0.421 196	.372 543
3	5/2	.494 581	.404 011	.049 877	.000 000	−0.183 710	−0.443 339	.536 377
3	7/2	.098 327	.392 102	.873 283	1.000 000	.828 156	.438 608	−0.201 365
5	9/2	.376 972	.374 135	−0.008 598	.000 000	.300 353	.517 102	−0.527 180
5	11/2	−0.356 890	−0.297 009	−0.219 175	.000 000	.209 080	.306 737	−0.382 637

K = 1/2, (530)

L	J							
E		6.201 605	6.186 085	6.184 555	6.125 000	5.983 786	5.823 125	5.671 473
A		.075 893	1.097 212	4.651 686	5.000 000	3.707 986	.629 178	−0.308 994
1	1/2	−0.085 906	.044 840	.136 815	.000 000	.049 090	−0.012 262	−0.077 938
1	3/2	−0.487 193	−0.371 160	.004 615	.000 000	.101 876	.353 134	.461 147
3	5/2	.357 909	.387 996	.361 688	.000 000	−0.292 479	−0.279 978	−0.242 251
3	7/2	−0.376 083	−0.463 541	−0.035 929	.000 000	−0.331 954	−0.539 452	−0.481 347
5	9/2	.575 171	.656 955	.921 492	1.000 000	.887 249	.663 449	.588 241
5	11/2	.393 579	.251 452	−0.001 844	.000 000	−0.065 292	−0.256 097	−0.380 606

K = 1/2, (521)

L	J							
E		6.533 152	6.440 255	6.408 249	6.405 000	6.256 575	6.151 957	6.084 592
A		.618 137	.867 223	.779 415	−2.000 000	.774 753	.946 972	.892 163
1	1/2	−0.568 830	−0.645 232	.711 406	.000 000	−0.479 054	−0.508 181	−0.498 886
1	3/2	.098 784	−0.066 144	.375 361	1.000 000	.501 268	.286 178	.156 120
3	5/2	−0.117 276	−0.208 651	.421 415	.000 000	.588 346	.485 940	.426 458
3	7/2	.486 569	.472 120	−0.303 067	.000 000	.304 052	.443 299	.480 818
5	9/2	.581 288	.530 681	−0.284 625	.000 000	.280 094	.456 606	.518 151
5	11/2	−0.279 778	−0.176 767	.050 532	.000 000	.046 767	.136 797	.212 618

K = 1/2, (510)

L	J							
E		6.915 252	6.708 742	6.528 839	6.430 000	6.392 558	6.417 308	6.460 968
A		−0.047 783	−0.007 903	.148 122	3.000 000	.158 453	−0.169 748	−0.340 257
1	1/2	−0.093 309	−0.054 446	.026 100	.000 000	−0.003 597	.058 870	.096 683
1	3/2	−0.530 813	−0.607 889	−0.695 401	.000 000	−0.709 728	−0.675 959	−0.635 723
3	5/2	.550 389	.567 905	.620 036	1.000 000	.631 999	.568 662	.540 451
3	7/2	.488 946	.427 659	.275 041	.000 000	−0.237 492	−0.371 500	−0.437 754
5	9/2	−0.376 555	−0.333 684	−0.233 106	.000 000	.199 065	.268 311	.296 294
5	11/2	−0.160 438	−0.103 697	−0.036 129	.000 000	−0.028 648	−0.078 960	−0.122 484

K = 1/2, (501)

L	J							
E		7.360 356	7.047 593	6.750 869	6.555 000	6.658 123	6.805 902	6.961 856
A		.497 631	.711 559	.959 258	1.000 000	.959 415	.852 191	.757 632
1	1/2	−0.526 998	−0.572 387	.674 720	1.000 000	.868 997	.815 398	.787 561

TABLE 3.4 (*contd.*)

N = 5 KAPPA = 0.0500 MU = 0.4500

L	J							
1	3/2	.528 630	.510 973	−0.454 269	.000 000	.315 639	.376 544	.403 536
3	5/2	.553 404	.565 440	−0.551 941	.000 000	.367 533	.405 166	.413 878
3	7/2	−0.291 794	−0.233 091	.134 739	.000 000	.075 832	.134 161	.171 423
5	9/2	−0.215 144	−0.187 721	.123 960	.000 000	.065 773	.103 351	.122 131
5	11/2	.071 676	.044 569	−0.015 257	.000 000	.007 706	.022 372	.036 192

N = 5 KAPPA = 0.0500 MU = 0.6300

DELTA =		−0.3	−0.2	−0.1	0.0	0.1	0.2	0.3
K = 11/2, (505)								
E		4.805 000	4.971 667	5.138 333	5.305 000	5.471 667	5.638 333	5.805 000
L	J							
5	11/2	1.000 000	1.000 000	1.000 000	1.000 000	1.000 000	1.000 000	1.000 000
K = 9/2, (514)								
E		5.055 000	5.145 110	5.227 484	5.305 000	5.379 452	5.451 913	5.523 029
L	J							
5	9/2	.254 824	.144 135	.061 060	.000 000	−0.045 360	−0.079 796	−0.106 565
5	11/2	.966 988	.989 558	.998 134	1.000 000	.998 971	.996 811	.994 306
K = 9/2, (505)								
E		5.405 000	5.548 223	5.699 183	5.855 000	6.013 881	6.174 754	6.336 971
L	J							
5	9/2	.966 988	.989 558	.998 134	1.000 000	.998 971	.996 811	.994 306
5	11/2	−0.254 824	−0.144 135	−0.061 060	.000 000	.045 360	.079 796	.106 565
K = 7/2, (523)								
E		5.203 933	5.263 793	5.294 453	5.305 000	5.302 379	5.291 131	5.274 142
L	J							
3	7/2	.434 294	.249 719	.103 250	.000 000	−0.071 294	−0.121 250	−0.157 170
5	9/2	.092 640	.091 434	.053 203	.000 000	−0.052 535	−0.098 309	−0.136 320
5	11/2	.895 995	.963 992	.993 232	1.000 000	.996 071	.987 742	.978 118
K = 7/2, (514)								
E		5.545 788	5.663 990	5.785 474	5.855 000	5.906 367	5.959 816	6.016 154
L	J							
3	7/2	.754 228	.789 502	−0.528 089	.000 000	.110 269	.132 161	.133 018
5	9/2	−0.581 223	−0.595 631	.849 137	1.000 000	.992 076	.984 625	.978 480
5	11/2	−0.305 484	−0.148 023	.009 412	.000 000	.060 217	.114 222	.157 744
K = 7/2, (503)								
E		5.782 279	5.804 216	5.852 073	5.972 000	6.123 254	6.281 053	6.441 704
L	J							
3	7/2	.492 473	.560 649	.842 889	1.000 000	.991 341	.983 785	.978 572
5	9/2	.808 454	.798 038	.525 486	.000 000	−0.114 129	−0.144 390	−0.154 900
5	11/2	−0.322 293	−0.220 927	−0.115 769	.000 000	.064 936	.106 393	.135 654

TABLE 3.4 (*contd.*)

N = 5 KAPPA = 0.0500 MU = 0.6300

K = 5/2, (532)

L	J							
E		5.327 950	5.351 636	5.343 169	5.305 000	5.242 059	5.160 171	5.064 568
3	5/2	.136 163	.053 975	.011 526	.000 000	.007 616	.024 166	.043 127
3	7/2	.420 859	.289 014	.140 526	.000 000	−0.116 900	−0.206 758	−0.273 278
5	9/2	.102 394	.071 762	.040 332	.000 000	−0.047 767	−0.096 426	−0.140 732
5	11/2	.890 984	.953 104	.989 188	1.000 000	.991 965	.973 329	.950 607

K = 5/2, (523)

L	J							
E		5.746 911	5.834 273	5.864 582	5.855 000	5.818 421	5.773 359	5.728 356
3	5/2	.702 987	.460 616	.167 617	.000 000	−0.118 245	−0.186 757	−0.226 332
3	7/2	.021 806	−0.094 307	−0.197 487	.000 000	.144 342	.180 763	.175 313
5	9/2	.683 242	.880 269	.965 778	1.000 000	.980 277	.955 761	.937 164
5	11/2	−0.196 252	−0.063 765	−0.013 275	.000 000	.065 122	.137 721	.199 408

K = 5/2, (512)

L	J							
E		5.911 569	5.925 181	5.955 741	5.972 000	6.001 951	6.045 694	6.097 096
3	5/2	.344 592	.431 977	.239 985	.000 000	−0.086 447	−0.123 152	−0.146 638
3	7/2	.732 088	.846 622	.948 258	1.000 000	.977 127	.948 207	.925 591
5	9/2	−0.476 635	−0.154 862	.150 279	.000 000	−0.161 483	−0.229 586	−0.258 454
5	11/2	−0.343 689	−0.269 528	−0.143 635	.000 000	.108 039	.181 734	.234 476

K = 1/2, (550)

L	J							
E		5.439 559	5.435 680	5.390 752	5.305 000	5.178 770	5.012 705	4.808 433
A		−5.385 965	−5.743 682	−5.937 787	−6.000 000	−5.937 865	−5.744 949	−5.404 103
1	1/2	.076 501	.023 241	.002 867	.000 000	−0.002 504	−0.017 618	−0.049 831
1	3/2	.185 288	.089 507	.023 340	.000 000	0.22 516	.083 109	.165 134
3	5/2	.038 070	.013 188	.003 074	.000 000	.004 629	.023 391	.061 145
3	7/2	.455 544	.328 971	.171 933	.000 000	−0.170 603	−0.323 735	−0.445 208
5	9/2	.026 383	.015 023	.008 365	.000 000	−0.012 284	−0.031 214	−0.059 825
5	11/2	.866 112	.939 589	.984 788	1.000 000	.984 992	.941 518	.874 483

K = 1/2, (541)

L	J							
E		5.880 769	5.943 534	5.930 982	5.855 000	5.720 990	5.527 783	5.291 831
A		2.648 244	4.219 950	4.862 987	5.000 000	4.795 253	3.871 360	2.910 898
1	1/2	.469 156	.247 500	.053 930	.000 000	.046 958	.158 107	.247 548
1	3/2	.230 585	.087 236	−0.007 775	.000 000	−0.042 019	−0.169 528	−0.250 406
3	5/2	.536 672	.452 854	.242 743	.000 000	−0.239 370	−0.425 823	−0.520 015
3	7/2	.112 849	.063 563	−0.027 733	.000 000	.081 720	.184 511	.170 576
5	9/2	.623 567	.847 828	.968 153	1.000 000	.965 011	.846 458	.722 319
5	11/2	−0.192 706	−0.056 599	−0.004 112	.000 000	.028 393	.120 008	.234 009

K = 1/2, (530)

E		6.046 947	6.050 207	6.042 852	5.972 000	5.840 530	5.676 886	5.517 018
A		−0.894 284	−2.703 571	−3.768 090	−4.000 000	−3.711 299	−2.604 722	−1.638 503

TABLE 3.4 (*contd.*)

N = 5 KAPPA = 0.0500 MU = 0.6300

L	J							
1	1/2	−0.289 642	.289 096	.096 790	.000 000	.064 402	.141 404	.168 576
1	3/2	−0.539 235	.533 280	.335 150	.000 000	−0.312 682	−0.472 100	−0.517 105
3	5/2	.135 519	.036 108	.048 721	.000 000	−0.047 126	−0.021 281	.043 765
3	7/2	−0.471 346	.698 450	.920 435	1.000 000	.925 049	.747 825	.589 192
5	9/2	.499 613	−0.230 896	.010 732	.000 000	−0.111 669	−0.335 916	−0.465 291
5	11/2	.367 678	−0.299 310	−0.169 166	.000 000	.166 361	.290 846	.372 326

K = 1/2, (521)

E	6.398 805	6.334 454	6.341 380	6.322 000	6.172 961	6.043 102	5.956 490
A	.447 062	.933 324	1.755 457	3.000 000	1.871 433	1.363 818	1.087 855

L	J							
1	1/2	−0.594 581	.673 906	.633 024	.000 000	−0.415 034	−0.482 186	−0.482 963
1	3/2	.102 278	.070 523	.240 790	.000 000	.339 112	.242 825	.136 318
3	5/2	−0.288 135	.430 161	.680 691	1.000 000	.790 430	.644 563	.553 874
3	7/2	.503 709	−0.430 289	−0.183 402	.000 000	.204 884	.387 968	.460 270
5	9/2	.493 217	−0.392 977	−0.209 147	.000 000	.212 923	.364 349	.446 813
5	11/2	−0.236 655	.127 512	.024 122	.000 000	.025 621	.099 009	.172 906

K = 1/2, (510)

E	6.812 454	6.621 325	6.464 839	6.387 000	6.335 506	6.346 052	6.380 036
A	−0.013 559	−0.039 644	−0.430 813	−2.000 000	−0.827 513	−0.595 717	−0.594 492

L	J							
1	1/2	−0.180 292	−0.186 486	.249 724	.000 000	−0.168 742	−0.165 144	−0.172 532
1	3/2	−0.535 314	−0.615 371	.751 667	1.000 000	.818 917	.724 019	.668 097
3	5/2	.592 974	.601 064	−0.534 827	.000 000	−0.478 985	−0.526 634	−0.525 809
3	7/2	.468 993	.401 819	−0.267 727	.000 000	.248 014	.367 264	.431 574
5	9/2	−0.302 744	−0.238 895	.118 676	.000 000	−0.096 677	−0.178 555	−0.223 797
5	11/2	−0.133 071	−0.082 068	.028 862	.000 000	.024 853	.066 445	.105 181

K = 5/2, (503)

E	6.267 569	6.209 577	6.223 842	6.322 000	6.458 236	6.608 109	6.763 980

L	J							
3	5/2	.607 063	.773 508	.956 127	1.000 000	.989 185	.974 357	.961 983
3	7/2	0.535 210	−0.436 817	−0.205 082	.000 000	.103 547	.159 622	.194 589
5	9/2	−0.543 613	−0.442 713	−0.207 514	.000 000	.103 436	.156 566	.187 405
5	11/2	.222 508	.121 987	.026 455	.000 000	.009 589	.025 227	.040 041

K = 3/2, (541)

E	5.400 790	5.407 718	5.375 001	5.305 000	5.200 274	5.064 765	4.903 884

L	J							
1	3/2	.195 482	.081 761	.018 412	.000 000	.013 391	.043 617	.078 484
3	5/2	.024 099	.022 346	.007 768	.000 000	.009 304	.035 379	.071 258
3	7/2	.451 691	.318 019	.162 054	.000 000	−0.151 387	−0.278 453	−0.375 535
5	9/2	.037 208	.039 286	.024 694	.000 000	−0.033 643	−0.074 792	−0.118 484
5	11/2	.869 366	.943 470	.986 270	1.000 000	.987 767	.955 885	.913 070

K = 3/2, (532)

E	5.816 060	5.897 258	5.908 716	5.855 000	5.754 598	5.623 512	5.481 435

TABLE 3.4 (*contd.*)

N = 5 KAPPA = 0.0500 MU = 0.6300

L	J							
1	3/2	.577 652	−0.368 457	−0.074 793	.000 000	−0.040 295	−0.094 732	−0.122 833
3	5/2	−0.451 088	.390 104	.219 756	.000 000	−0.193 681	−0.323 909	−0.397 806
3	7/2	.266 923	−0.290 917	−0.131 167	.000 000	.144 750	.204 411	.188 572
5	9/2	−0.581 504	.787 201	.963 795	1.000 000	.967 783	.907 065	.859 018
5	11/2	−0.231 180	.087 972	−0.002 914	.000 000	.057 518	.146 829	.230 631

K = 3/2, (521)

E	6.002 460	6.005 892	6.012 955	5.972 000	5.904 065	5.840 473	5.786 021

L	J							
1	3/2	.519 489	.547 999	.327 385	.000 000	−0.191 149	−0.282 289	−0.330 616
3	5/2	.302 420	.169 053	.079 854	.000 000	−0.083 497	−0.106 882	−0.115 785
3	7/2	.390 174	.632 002	.918 264	1.000 000	.951 412	.869 528	.801 920
5	9/2	.601 144	.437 882	.131 682	.000 000	−0.175 482	−0.303 484	−0.362 896
5	11/2	−0.353 642	−0.282 758	−0.160 917	.000 000	.143 216	.246 388	.320 184

K = 3/2, (512)

E	6.351 680	6.278 216	6.276 880	6.322 000	6.302 773	6.318 119	6.353 675

L	J							
1	3/2	−0.394 891	.560 118	.736 732	.000 000	.269 627	.257 659	.230 158
3	5/2	.509 562	−0.547 040	−0.613 758	1.000 000	.932 798	.884 554	.849 418
3	7/2	.595 052	−0.512 715	−0.228 806	.000 000	.164 965	.271 633	.329 660
5	9/2	−0.428 479	.329 979	.166 051	.000 000	.172 120	.272 508	.329 144
5	11/2	−0.216 161	.123 499	.024 518	.000 000	.018 704	.055 954	.092 222

K = 3/2, (501)

E	6.770 010	6.585 249	6.434 115	6.387 000	6.512 624	6.660 797	6.815 986

L	J							
1	3/2	.449 774	.493 477	.586 611	1.000 000	.942 847	.918 177	.903 589
3	5/2	.666 954	.720 755	.754 036	.000 000	−0.292 091	−0.316 184	−0.318 992
3	7/2	−0.467 337	−0.389 780	−0.246 928	.000 000	.154 046	.225 423	.267 699
5	9/2	−0.339 854	−0.279 543	−0.159 928	.000 000	−0.042 959	−0.072 638	−0.089 554
5	11/2	.137 735	.083 189	.027 687	.000 000	.012 116	.029 789	.045 707

K = 1/2, (501)

E	7.299 467	6.992 800	6.707 195	6.537 000	6.629 243	6.771 472	6.924 191
A	.198 502	.333 624	.518 245	1.000 000	.809 990	.710 209	.638 344

L	J							
1	1/2	−0.551 467	−0.604 732	.724 316	1.000 000	.890 457	.833 619	.802 993
1	3/2	.569 827	.562 437	−0.513 894	.000 000	.338 136	.397 878	.421 538
3	5/2	.507 473	.497 121	−0.435 100	.000 000	.293 689	.353 369	.375 071
3	7/2	−0.289 657	−0.230 865	.130 978	.000 000	.070 800	.128 479	.165 654
5	9/2	−0.161 053	−0.127 070	.068 318	.000 000	.038 074	.071 154	.091 942
5	11/2	.061 755	.037 265	−0.012 019	.000 000	.005 892	.018 234	.030 557

N = 6 KAPPA = 0.0500, MU = 0.4480

DELTA =	−0.3	−0.2	−0.1	0.0	0.1	0.2	0.3

K = 13/2, (606)

E	5.659 200	5.859 200	6.059 200	6.259 200	6.459 200	6.659 200	6.859 200

TABLE 3.4 (*contd.*)

N = 6 KAPPA = 0.0500 MU = 0.4480

L	J							
6	13/2	1.000 000	1.000 000	1.000 000	1.000 000	1.000 000	1.000 000	1.000 000
K = 11/2, (615)								
E		5.920 600	6.037 979	6.150 255	6.259 200	6.365 927	6.471 132	6.575 261
L	J							
6	11/2	.190 623	.108 871	.046 975	.000 000	−0.036 202	−0.064 645	−0.087 428
6	13/2	.981 663	.994 056	.998 896	1.000 000	.999 345	.997 908	.996 171
K = 11/2, (606)								
E		6.347 800	6.530 421	6.718 145	6.909 200	7.102 473	7.297 268	7.493 139
L	J							
6	11/2	.981 663	.994 056	.998 896	1.000 000	.999 345	.997 908	.996 171
6	13/2	−0.190 623	−0.108 871	−0.046 974	.000 000	.036 202	.064 645	.087 428
K = 9/2, (624)								
E		6.060 872	6.160 091	6.221 000	6.259 200	6.284 873	6.303 171	6.316 804
L	J							
4	9/2	.528 243	.296 541	.114 619	.000 000	−0.071 681	−0.118 752	−0.151 291
6	11/2	.066 924	.074 593	.044 460	.000 000	−0.043 601	−0.081 680	−0.113 727
6	13/2	.846 452	.952 103	.992 414	1.000 000	.996 474	.989 559	.981 925
K = 9/2, (615)								
E		6.370 754	6.506 715	6.672 008	6.852 000	6.989 342	7.079 601	7.171 083
L	J							
4	9/2	.806 821	.928 942	.980 044	1.000 000	.385 680	.236 164	.190 835
6	11/2	−0.350 188	−0.253 914	−0.168 384	.000 000	.920 123	.965 687	.971 312
6	13/2	−0.475 823	−0.269 434	−0.105 646	.000 000	.068 004	.108 050	.141 901
K = 9/2, (604)								
E		6.688 774	6.753 594	6.827 392	6.909 200	7.046 185	7.237 628	7.432 513
L	J							
4	9/2	.264 573	.221 655	.162 410	.000 000	919 844	.964 430	969 893
6	11/2	.934 286	.964 346	.984 718	1.000 000	−0.389 195	−0.246 529	−0.208 854
6	13/2	−0.238 980	−0.144 588	−0.062 873	.000 000	.049 140	.095 387	.125 247
K = 5/2, (642)								
E		6.289 847	6.327 891	6.315 784	6.259 200	6.164 425	6.039 785	5.893 712
L	J							
2	5/2	.260 914	.103 849	.021 538	.000 000	.012 849	.038 338	.064 607
4	7/2	.035 090	.029 208	.009 322	.000 000	.009 103	.031 493	.059 112
4	9/2	.511 909	.366 347	.185 086	.000 000	−0.160 775	−0.283 423	−0.370 208
6	11/2	.039 534	.042 881	.027 403	.000 000	−0.035 996	−0.075 954	−0.114 737
6	13/2	.816 749	.923 172	.982 060	1.000 000	.986 209	.954 694	.917 667
K = 5/2, (633)								
E		6.668 308	6.792 431	6.862 245	6.852 000	6.800 837	6.705 429	6.599 135

TABLE 3.4 (*contd.*)

N = 6 KAPPA = 0.0500 MU = 0.4480

L	J							
2	5/2	.763 135	.688 460	.334 917	.000 000	−0.136 867	−0.140 974	−0.135 414
4	7/2	−0.300 683	−0.166 866	.001 733	.000 000	−0.177 337	−0.312 498	−0.376 322
4	9/2	.284 035	.579 490	.905 025	1.000 000	.767 006	.450 778	.311 056
6	11/2	−0.301 796	−0.280 649	−0.197 561	.000 000	.582 328	.796 104	.821 381
6	13/2	−0.394 283	−0.289 149	−0.172 416	.000 000	.149 715	.213 133	.261 960

K = 5/2, (622)

E		6.902 298	6.932 687	6.945 157	6.909 200	6.850 978	6.822 912	6.807 484
L	**J**							
2	5/2	.252 446	.116 456	.017 761	.000 000	.093 696	−0.207 110	−0.265 683
4	7/2	.564 642	.498 288	.285 663	.000 000	−0.096 543	.027 284	0.020 311
4	9/2	.244 899	.294 826	.185 378	.000 000	−0.592 885	.796 242	.802 889
6	11/2	.687 318	.786 100	937 867	1.000 000	.791 043	−0.530 782	−0.448 942
6	13/2	−0.291 666	−0.182 405	−0.064 209	.000 000	−0.068 111	.201 572	.287 785

K = 5/2, (613)

E		7.163 604	7.109 023	7.136 180	7.265 600	7.327 511	7.390 642	7.467 760
L	**J**							
2	5/2	−0.436 904	.633 677	.891 583	1.000 000	.396 148	.308 014	.260 234
4	7/2	.315 628	−0.310 377	−0.314 711	.000 000	.890 487	.885 698	.869 848
4	9/2	.661 569	−0.589 903	−0.292 366	.000 000	.143 685	.217 003	.261 429
6	11/2	−0.447 831	.360 294	.139 134	.000 000	.170 851	.267 633	.319 754
6	13/2	−0.266 961	.155 953	.034 653	.000 000	.016 280	.044 129	.071 093

K = 5/2, (602)

E		7.563 943	7.425 968	7.328 634	7.302 000	7.444 250	7.629 232	7.819 910
L	**J**							
2	5/2	.308 117	.316 433	.303 531	.000 000	.902 990	.917 000	.916 067
4	7/2	.699 941	.791 630	.905 132	1.000 000	−0.407 654	−0.340 712	−0.312 972
4	9/2	−0.399 531	−0.308 233	−0.163 799	.000 000	.117 026	.188 095	.230 683
6	11/2	−0.484 145	−0.414 285	−0.247 521	.000 000	−0.068 257	−0.084 212	−0.091 530
6	13/2	.145 345	.080 949	.022 529	.000 000	.008 585	.023 555	.037 273

K = 7/2, (633)

E		6.204 260	6.258 437	6.275 715	6.259 200	6.217 245	6.158 132	6.087 748
L	**J**							
4	7/2	.130 374	.051 235	.010 533	.000 000	.006 227	.018 996	.033 084
4	9/2	.482 779	.332 567	.157 730	.000 000	−0.120 022	−0.205 104	−0.264 797
6	11/2	.085 542	.063 782	.037 136	.000 000	−0.042 852	−0.084 792	−0.122 268
6	13/2	.861 748	.939 524	.986 728	1.000 000	.991 826	.974 875	.955 949

K = 7/2, (624)

E		6.661 537	6.755 590	6.795 531	6.852 000	6.888 009	6.881 227	6.872 011
L	**J**							
4	7/2	.809 546	.246 713	.086 575	.000 000	−0.127 524	−0.188 220	−0.219 840
4	9/2	.185 782	.894 767	.965 939	1.000 000	.604 699	.354 646	.268 505
5	11/2	.484 448	−0.195 138	−0.193 782	.000 000	.778 778	.902 344	.916 457
6	13/2	−0.274 646	−0.316 931	−0.148 038	.000 000	.107 623	.156 765	.199 201

K = 7/2, (613)

E		6.742 483	6.811 134	6.893 759	6.909 200	6.938 791	7.017 967	7.106 320

TABLE 3.4 (*contd.*)

N = 6 KAPPA = 0.0500 MU = 0.4480

L	J							
4	7/2	.011 390	.607 460	.270 232	.000 000	.008 536	−0.046 990	−0.080 125
4	9/2	.762 817	−0.023 026	.156 209	.000 000	.784 082	.904 901	.914 717
6	11/2	−0.525 262	.790 115	.947 912	1.000 000	−0.616 842	−0.392 752	−0.333 617
6	13/2	−0.376 936	−0.078 615	−0.063 530	.000 000	.068 178	.157 137	.213 478

K = 7/2, (604)

	E	7.114 120	7.097 239	7.157 396	7.302 000	7.478 355	7.665 075	7.856 322
L	**J**							
4	7/2	−0.572 284	.753 326	.958 837	1.000 000	.991 779	.980 818	.971 677
4	9/2	.387 970	0.297 087	−0.132 974	.000 000	.071 758	.115 382	.145 193
2	11/2	.694 329	−0.577 556	−0.250 064	.000 000	.105 714	.155 986	.183 999
6	13/2	−0.199 696	.103 289	.020 432	.000 000	.007 025	.018 731	.030 124

K = 3/2, (651)

	E	6.355 277	6.374 180	6.342 113	6.259 200	6.127 914	5.953 335	.5.743 451
L	**J**							
2	3/2	.092 167	.027 090	.003 108	.000 000	−0.002 110	−0.012 658	−0.030 772
2	5/2	.236 295	.116 798	.029 892	.000 000	.024 825	.082 062	.146 580
4	7/2	.051 537	.022 862	.006 462	.000 000	.008 436	.034 364	.072 504
4	9/2	.514 026	.382 175	.201 404	.000 000	−0.191 250	−0.347 167	−0.457 661
6	11/2	.036 126	.027 065	.016 757	.000 000	−0.024 116	−0.054 988	−0.089 379
6	13/2	.816 998	.915 594	.978 882	1.000 000	.980 892	.931 867	.868 831

K = 3/2, (642)

	E	6.815 964	6.912 744	6.912 254	6.852 000	6.732 972	6.558 232	6.357 375
L	**J**							
2	3/2	−0.637 341	.327 112	.060 595	.000 000	.047 795	.124 238	.169 457
2	5/2	−0.335 350	.558 192	.347 754	.000 000	−0.261 308	−0.304 603	−0.282 854
4	7/2	−0.468 509	.150 542	.005 778	.000 000	−0.165 078	−0.381 980	−0.484 920
4	9/2	−0.169 425	.654 149	.905 065	1.000 000	.856 726	.517 620	.309 006
6	11/2	−0.360 688	.044 319	−0.135 244	.000 000	.366 440	.635 184	.673 931
6	13/2	.320 990	−0.358 999	−0.194 750	.000 000	.184 186	.272 918	.326 288

K = 3/2, (631)

	E	6.927 575	6.955 197	6.975 263	6.909 200	6.785 834	6.658 679	6.541 981
L	**J**							
2	3/2	.010 483	.280 174	.089 570	.000 000	.019 568	0.008 708	.047 164
2	5/2	.519 719	−0.119 421	.034 761	.000 000	.097 651	.303 315	−0.414 105
4	7/2	−0.363 253	.506 608	.304 191	.000 000	−0.205 293	−0.178 507	.129 078
4	9/2	.410 562	−0.182 075	.110 164	.000 000	−0.381 841	−0.635 448	.631 227
6	11/2	−0.545 588	.784 359	.940 386	1.000 000	.894 042	.651 542	−0.545 446
6	13/2	−0.362 769	.047 108	−0.042 118	.000 000	−0.053 133	−0.218 536	.337 153

K = 3/2, (622)

	E	7.262 900	7.235 150	7.283 682	7.265 600	7.191 668	7.129 596	7.093 170
L	**J**							
2	3/2	−0.631 999	.750 295	.686 188	.000 000	−0.312 280	−0.387 613	−0.411 914
2	5/2	.132 981	.072 271	.533 919	1.000 000	.533 149	.363 420	.260 700
4	7/2	−0.011 962	.200 110	.373 390	.000 000	.723 212	.687 624	.641 121
4	9/2	.407 845	−0.376 258	−0.271 108	.000 000	.222 839	.328 242	.377 933
6	11/2	.595 187	−0.481 693	−0.172 424	.000 000	.211 522	.361 324	.437 122
6	13/2	−0.249 329	.134 876	.037 784	.000 000	.028 263	.080 982	.131 973

TABLE 3.4 (*contd.*)

N = 6 KAPPA = 0.0500 MU = 0.4480

K = 3/2, (611)

E		7.602 204	7.441 913	7.324 861	7.302 000	7.301 021	7.359 594	7.435 264
L	J							
2	3/2	−0.072 614	−0.024 938	.204 100	.000 000	.010 576	0.044 599	−0.080 01
2	5/2	−0.570 875	−0.675 201	−0.676 644	.000 000	.778 212	.787 121	.765 48
4	7/2	.486 301	.495 974	.638 373	1.000 000	−0.584 253	−0.496 636	−0.462 77
4	9/2	.519 580	.440 219	.203 720	.000 000	.180 869	.307 666	.378 95
6	11/2	−0.366 231	−0.306 029	−0.225 788	.000 000	−0.141 122	−0.185 565	−0.206 497
6	13/2	−0.168 082	−0.100 219	−0.022 250	.000 000	.017 147	.052 064	.085 012

K = 3/2, (602)

E		8.039 680	7.784 417	7.565 426	7.515 600	7.664 191	7.844 165	8.032 359
L	J							
2	3/2	−0.424 847	−0.500 201	.689 772	1.000 000	.948 524	.912 193	.889 965
2	5/2	.466 920	.446 540	−0.366 127	.000 000	.178 058	.238 430	.270 354
4	7/2	.639 734	.658 884	−0.600 378	.000 000	.257 187	.318 704	.343 134
4	9/2	−0.319 702	−0.239 827	.114 699	.000 000	.035 631	.073 139	.100 878
6	11/2	−0.287 332	−0.237 490	.128 029	.000 000	.034 250	.063 409	.081 248
6	13/2	.086 380	.048 511	−0.012 837	.000 000	.003 111	.010 630	.018 771

K = 1/2, (660)

E		6.383 718	6.396 760	6.355 165	6.259 200	6.109 164	5.905 961	5.652 393
A		6.100 450	6.619 001	6.907 259	7.000 000	6.907 879	6.626 627	6.150 146
L	J							
0	1/2	.124 198	.039 482	.004 973	.000 000	−0.004 246	−0.028 587	−0.075 894
2	3/2	−0.018 136	−0.000 264	.000 681	.000 000	−0.001 786	−0.015 887	−0.051 534
2	5/2	.257 508	.128 119	.034 105	.000 000	.033 342	.121 287	.233 345
4	7/2	−0.002 846	.005 057	.002 200	.000 000	.003 598	.018 607	.050 813
4	9/2	.518 917	.390 417	.209 132	.000 000	−0.207 903	−0.385 698	−0.511 210
6	11/2	.003 491	.008 112	.005 616	.000 000	−0.008 539	−0.021 522	−0.041 621
6	13/2	.805 382	.910 774	.977 261	1.000 000	.977 526	.913 590	.819 441

K = 1/2, (651)

E		6.817 806	6.923 852	6.934 424	6.852 000	6.693 210	6.454 972	6.156 880
A		.552 808	2.590 871	4.707 568	5.000 000	4.505 060	1.363 768	−0.631 627
L	J							
0	1/2	.543 106	.404 284	.113 039	.000 000	.083 527	.215 767	.266 648
2	3/2	−0.354 534	−0.178 892	−0.006 940	.000 000	.051 365	.232 112	.361 424
2	5/2	.438 261	.540 205	.371 057	.000 000	−0.355 152	−0.474 815	−0.408 352
4	7/2	−0.360 655	−0.209 574	−0.012 737	.000 000	−0.081 485	−0.331 862	−0.474 533
4	9/2	.186 865	.543 408	.896 093	1.000 000	.892 233	.543 202	.241 704
6	11/2	−0.313 447	−0.263 034	−0.066 128	.000 000	.141 313	.408 433	.461 480
6	13/2	−0.352 178	−0.323 002	−0.204 873	.000 000	.203 867	.319 532	.367 361

K = 1/2, (640)

E		6.980 894	6.996 590	6.990 658	6.909 200	6.748 079	6.535 466	6.319 713
A		−1.449 552	−3.573 607	−5.722 667	−6.000 000	−5.515 407	−2.352 376	−0.247 070
L	J							
0	1/2	.188 232	.016 248	−0.017 694	.000 000	.002 394	−0.072 068	−0.174 713
2	3/2	.231 507	.229 246	.084 403	.000 000	.068 685	.105 276	.054 289
2	5/2	.380 951	.211 277	.015 810	.000 000	.045 472	.290 760	.449 363

TABLE 3.4 (*contd.*)

N = 6 KAPPA = 0.0500 MU = 0.4480

4	7/2	.479 841	.480 514	.309 314	.000 000	−0.292 936	−0.380 720	−0.349 132
4	9/2	.285 969	.299 146	.065 695	.000 000	−0.155 839	−0.428 689	−0.414 864
6	11/2	.586 898	.744 295	.944 401	1.000 000	.939 400	.729 664	.591 259
6	13/2	−0.330 719	−0.167 889	−0.020 703	.000 000	−0.025 275	−0.195 065	−0.347 862

K = 1/2, (631)

E		7.278 496	7.237 789	7.278 602	7.265 600	7.079 871	6.878 631	6.726 133
A		.037 012	.093 694	.714 226	3.000 000	−0.165 556	−0.877 259	−0.957 676
L	J							
0	1/2	−0.467 642	−0.599 286	.621 376	.000 000	−0.368 567	−0.373 444	−0.327 933
2	3/2	.415 202	.440 030	−0.361 078	.000 000	−0.444 337	−0.545 153	−0.545 266
2	5/2	.135 551	−0.103 693	.515 827	1.000 000	.549 635	.238 558	.053 257
4	7/2	.029 810	.159 903	−0.333 504	.000 000	.490 463	.366 970	.259 056
4	9/2	.491 535	.461 914	.278 558	.000 000	.281 915	.422 479	.459 934
6	11/2	−0.521 303	−0.416 747	.164 728	.000 000	.207 576	.420 605	.516 196
6	13/2	−0.276 211	−0.154 489	.038 503	.000 000	.038 806	.127 960	.217 256

K = 1/2, (620)

E		7.646 235	7.494 012	7.379 665	7.302 000	7.177 757	7.109 639	7.071 431
A		−0.697 287	−0.961 116	−1.592 840	−4.000 000	−0.772 723	.019 338	.281 823
L	J							
0	1/2	.378 013	.331 790	.180 561	.000 000	.217 811	.340 993	.390 707
2	3/2	.416 450	.433 611	.423 527	.000 000	−0.232 866	−0.193 746	−0.165 383
2	5/2	.231 453	.359 896	.473 842	.000 000	−0.603 722	−0.550 372	−0.463 927
4	7/2	.421 546	.506 311	.673 842	1.000 000	.663 450	.537 641	.487 020
4	9/2	−0.462 794	−0.392 758	−0.217 804	.000 000	−0.215 188	−0.387 621	−0.468 421
6	11/2	−0.454 186	−0.388 485	−0.247 319	.000 000	.216 204	.310 507	.355 970
6	13/2	.178 723	.104 126	.028 764	.000 000	−0.025 207	−0.086 914	−0.146 451

K = 1/2, (611)

E		8.060 490	7.794 873	7.563 989	7.500 000	7.442 848	7.486 760	7.556 850
A		.032 774	.002 960	−0.015 994	1.000 000	−0.952 747	−0.919 777	−0.840 366
L	J							
0	1/2	−0.285 444	−0.371 225	.559 613	1.000 000	.466 199	.351 812	.287 038
2	3/2	.321 883	.330 153	−0.389 464	.000 000	.660 087	.582 025	.548 114
2	5/2	.577 063	.592 080	−0.529 003	.000 000	.381 957	.461 497	.479 585
4	7/2	−0.534 906	−0.530 071	.472 302	.000 000	.430 428	.512 448	.530 669
4	9/2	−0.352 982	−0.277 763	.140 643	.000 000	.094 525	.191 155	.252 704
6	11/2	.250 738	.204 171	−0.110 647	.000 000	.082 216	.155 597	.195 474
6	13/2	.091 214	.053 092	−0.014 640	.000 000	.009 441	.033 792	.059 160

K = 1/2, (600)

E		8.535 963	8.159 725	7.801 097	7.515 600	7.652 671	7.832 171	8.020 201
A		−0.576 206	−0.771 804	−0.997 553	−2.000 000	−0.006 507	.139 679	.244 771
L	J							
0	1/2	.459 476	.477 175	.504 983	.000 000	.769 659	.753 608	.741 641
2	3/2	.609 056	.651 765	.728 967	1.000 000	−0.552 506	−0.511 181	−0.488 543
2	5/2	−0.437 661	−0.395 336	−0.300 757	.000 000	.241 281	.320 156	.358 894
4	7/2	−0.418 280	−0.402 821	−0.340 295	.000 000	−0.203 832	−0.241 340	−0.253 841
4	9/2	.183 432	.134 233	.064 436	.000 000	.041 151	.084 357	.115 390
6	11/2	.127 982	.102 407	.056 189	.000 000	−0.029 889	−0.052 687	−0.065 829
6	13/2	−0.037 426	−0.021 101	−0.005 928	.000 000	.003 345	.011 476	.020 148

N = 6 KAPPA = 0.0500 MU = 0.6200

DELTA =		−0.3	−0.2	−0.1	0.0	0.1	0.2	0.3
K = 13/2, (606)								
E		5.298 000	5.498 000	5.698 000	5.898 000	6.098 000	6.298 000	6.498 000
L	J							
6	13/2	1.000 000	1.000 000	1.000 000	1.000 000	1.000 000	1.000 000	1 000 000
K = 11/2, (615)								
E		5.559 400	5.676 779	5.789 055	5.898 000	6.004 727	6.109 932	6.214 061
L	J							
6	11/2	.190 623	.108 871	.046 975	.000 000	−0.036 202	−0.064 645	−0.087 428
6	13/2	.981 663	.994 056	.998 896	1.000 000	.999 345	.997 908	.996 171
K = 11/2, (606)								
E		5.986 600	6.169 221	6.356 945	6.548 000	6.741 273	6.936 068	7.131 939
L	J							
6	11/2	.981 663	.994 056	.998 896	1.000 000	.999 345	.997 908	.996 171
6	13/2	−0.190 623	−0.108 871	−0.046 974	.000 000	.036 202	.064 645	.087 428
K = 9/2, (624)								
E		5.735 542	5.809 964	5.861 559	5.898 000	5.924 450	5.944 190	5.959 311
L	J							
4	9/2	.351 777	.196 296	.081 085	.000 000	−0.057 317	−0.098 742	−0.129 481
6	11/2	.106 435	.086 666	.046 013	.000 000	−0.043 232	−0.080 925	−0.112 824
6	13/2	.930 013	.976 707	.995 645	1.000 000	.997 420	.991 817	.985 142
K = 9/2, (615)								
E		6.139 174	6.298 687	6.450 055	6.548 000	6.634 855	6.723 318	6.813 969
L	J							
4	9/2	.813 998	.808 580	−0.430 307	.000 000	.087 642	.109 116	.112 783
6	11/2	−0.525 375	−0.577 775	.902 658	1.000 000	.994 987	.989 797	.985 383
6	13/2	−0.247 768	−0.111 239	−0.006 671	.000 000	.048 163	.091 623	.127 675
K = 9/2, (604)								
E		6.351 284	6.417 349	6.514 386	6.680 000	6.866 694	7.058 491	7.252 720
L	J							
4	9/2	.462 235	.554 677	.899 033	1.000 000	.994 502	.989 113	.985 147
6	11/2	.844 188	.811 582	.427 892	.000 000	−0.090 176	−0.117 270	−0.127 639
6	13/2	−0.271 453	−0.183 492	−0.092 992	.000 000	.053 241	.088 905	.114 864
K = 5/2, (642)								
E		5.980 816	5.988 166	5.959 559	5.898 000	5.807 084	5.691 364	5.555 823
L	J							
2	5/2	.139 687	.054 959	.011 780	.000 000	.008 023	.025 850	.046 638
4	7/2	.041 911	.024 187	.006 766	.000 000	.006 530	.023 615	.046 517
4	9/2	.410 638	.280 640	.139 310	.000 000	−0.125 362	−0.229 581	−0.311 305
6	11/2	.059 667	.049 817	.028 861	.000 000	−0.034 996	−0.073 285	−0.111 267
6	13/2	.898 080	.956 637	.989 735	1.000 000	.991 440	.969 895	.941 472
K = 5/2, (633)								
E		6.498 028	6.579 329	6.588 210	6.548 000	6.473 268	6.377 900	6.274 144

TABLE 3.4 (*contd.*)

N = 6 KAPPA = 0.0500 MU = 0.6200

L	J							
2	5/2	−0.552 042	−0.257 198	−0.046 779	.000 000	−0.021 853	−0.052 462	−0.071 851
4	7/2	.396 275	.329 041	.170 978	.000 000	−0.144 939	−0.247 253	−0.313 352
4	9/2	−0.336 224	−0.289 154	−0.142 717	.000 000	.122 619	.172 777	.172 136
6	11/2	.626 868	.860 122	.973 720	1.000 000	.980 232	.944 425	.912 809
6	13/2	.179 459	.046 493	−0.008 918	.000 000	.051 237	.119 676	.183 840

K = 5/2, (622)

E		6.645 540	6.682 991	6.700 696	6.680 000	6.649 020	6.623 371	6.604 452
L	J							
2	5/2	.521 627	.498 722	.233 660	.000 000	−0.137 164	−0.214 029	−0.260 833
4	7/2	.325 160	.179 983	.095 719	.000 000	−0.065 640	−0.095 437	−0.112 981
4	9/2	.425 861	.730 062	.948 147	1.000 000	.971 029	.920 820	.874 934
6	11/2	.576 507	.340 037	.132 097	.000 000	−0.140 471	−0.231 752	−0.279 645
6	13/2	−0.329 331	−0.265 081	−0.140 743	.000 000	.119 365	.208 482	.274 757

K = 5/2, (613)

E		6.975 672	6.971 112	7.049 151	7.130 000	7.160 810	7.215 086	7.284 747
L	J							
2	5/2	.489 797	.688 433	.775 131	.000 000	.199 052	.208 164	.195 816
4	7/2	−0.491 957	−0.518 435	−0.602 531	1.000 000	.964 533	.933 665	.908 892
4	9/2	−0.584 831	−0.429 878	−0.145 416	.000 000	.111 173	.192 072	.243 792
6	11/2	.374 452	.255 002	.121 834	.000 000	.132 585	.216 609	.269 788
6	13/2	.189 306	.086 387	.011 809	.000 000	.010 774	.033 551	.057 889

K = 5/2, (602)

E		7.369 944	7.248 401	7.172 384	7.214 000	7.379 818	7.562 279	7.750 833
L	J							
2	5/2	.404 649	.456 251	.585 020	1.000 000	.970 063	.952 596	.941 427
4	7/2	.702 467	.768 100	.773 637	.000 000	−0.210 517	−0.239 728	−0.246 571
4	9/2	−0.441 526	−0.346 188	−0.200 241	.000 000	.118 287	.180 663	.220 261
6	11/2	−0.361 804	−0.277 592	−0.136 906	.000 000	−0.024 697	−0.045 403	−0.058 416
6	13/2	.130 200	.070 382	.019 925	.000 000	.007 621	.019 782	.031 475

K = 3/2, (651)

E		6.048 619	6.037 434	5.986 930	5.898 000	5.772 185	5.611 966	5.421 118
L	J							
2	3/2	.046 393	.013 013	.001 484	.000 000	−0.001 106	−0.007 237	−0.019 378
2	5/2	.149 752	.069 229	.017 202	.000 000	.014 981	.052 833	.101 680
4	7/2	.036 816	.017 076	.004 744	.000 000	.005 814	.024 169	.053 542
4	9/2	.424 361	.299 243	.153 347	.000 000	−0.147 565	−0.277 676	−0.382 706
6	11/2	.040 058	.030 357	.017 702	.000 000	−0.023 200	−0.051 802	−0.084 286
6	13/2	.890 157	.950 936	.987 852	1.000 000	.988 649	.957 489	.912 606

K = 3/2, (642)

E		6.612 250	6.649 669	6.624 799	6.548 000	6.423 493	6.258 465	6.067 901
L	J							
2	3/2	.374 940	.137 910	.029 360	.000 000	.022 149	.069 204	.113 549
2	5/2	.060 245	−0.048 897	−0.026 434	.000 000	−0.031 550	−0.097 910	−0.146 310
4	7/2	.532 219	.388 376	.199 969	.000 000	−0.188 190	−0.335 729	−0.430 955
4	9/2	.020 758	−0.081 391	−0.076 857	.000 000	.103 507	.174 393	.175 303
6	11/2	.750 341	.906 121	.975 965	1.000 000	.975 084	.910 739	.843 249
6	13/2	−0.095 350	−0.008 615	−0.006 102	.000 000	.039 940	.114 248	.195 391

TABLE 3.4 (*contd.*)

N = 6 KAPPA = 0.0500 MU = 0.6200

K = 3/2, (631)

	E	6.733 142	6.763 543	6.748 346	6.680 000	6.573 595	6.452 915	6.333 504
L	J							
2	3/2	.350 608	.224 922	.050 433	.000 000	.025 636	.063 001	.090 221
2	5/2	.580 257	.492 387	.273 395	.000 000	−0.226 992	−0.364 130	−0.437 144
4	7/2	.005 154	.117 443	.063 532	.000 000	−0.049 506	−0.057 887	−0.046 310
4	9/2	.564 485	.781 398	.943 887	1.000 000	.953 599	.852 679	.751 997
6	11/2	−0.288 359	.009 457	.066 246	.000 000	−0.124 572	−0.260 563	−0.346 021
6	13/2	−0.372 231	−0.287 228	−0.152 851	.000 000	.143 169	.255 213	.336 734

K = 3/2, (622)

	E	7.082 549	7.100 648	7.159 776	7.130 000	7.040 167	6.963 667	6.912 689
L	J							
2	3/2	−0.694 440	.753 031	.492 338	.000 000	−0.250 170	−0.340 337	−0.376 018
2	5/2	.128 298	.055 331	−0.010 000	.000 000	.263 426	.254 043	.202 972
4	7/2	−0.249 042	.473 554	.845 997	1.000 000	.904 926	.812 585	.745 054
4	9/2	.416 678	−0.289 592	−0.065 318	.000 000	.136 865	.262 628	.335 632
6	11/2	.477 020	−0.339 856	−0.193 509	.000 000	.173 716	.295 523	.373 460
6	13/2	−0.195 198	.079 142	.008 979	.000 000	.014 912	.055 050	.100 930

K = 7/2, (633)

	E	5.879 896	5.913 190	5.918 005	5.898 000	5.858 203	5.803 522	5.737 972
L	J							
4	7/2	.088 376	.033 869	.007 065	.000 000	.004 631	.014 879	.027 030
4	9/2	.376 116	.247 257	.116 372	.000 000	−0.094 744	−0.168 330	−0.224 457
6	11/2	.096 733	.070 622	.038 797	.000 000	−0.042 093	−0.083 071	−0.120 182
6	13/2	.917 262	.965 779	.992 422	1.000 000	.994 601	.982 111	.966 667

K = 7/2, (624)

	E	6.401 508	6.488 098	6.532 131	6.548 000	6.544 701	6.535 581	6.526 111
L	J							
4	7/2	.601 541	.291 093	.117 407	.000 000	−0.087 782	−0.143 356	−0.178 769
4	9/2	−0.067 205	−0.273 963	−0.216 372	.000 000	.115 378	.149 817	.152 004
6	11/2	.787 878	.916 602	.969 134	1.000 000	.988 003	.972 054	.958 902
6	13/2	−0.113 489	−0.007 094	−0.013 351	.000 000	.053 213	.110 071	.159 511

K = 7/2, (613)

	E	6.511 204	6.572 860	6.627 544	6.680 000	6.748 339	6.827 578	6.913 114
L	J							
4	7/2	.407 826	.405 110	.158 235	.000 000	−0.062 919	−0.095 322	−0.117 365
4	9/2	.797 503	.870 770	.960 598	1.000 000	.986 092	.967 114	.951 055
6	11/2	−0.291 680	.129 701	.193 625	.000 000	−0.125 534	−0.180 320	−0.205 668
6	13/2	−0.335 543	−0.246 624	−0.121 337	0.000 000	.088 913	.151 952	.198 544

K = 7/2, (604)

	E	6.863 391	6.881 853	6.978 319	7.130 000	7.304 757	7.489 320	7.678 803
L	J							
4	7/2	.681 187	.866 028	.980 371	1.000 000	.994 140	.984 958	.976 492
4	9/2	−0.466 914	−0.324 912	−0.129 970	.000 000	.073 039	.117 943	.148 349
6	11/2	−0.533 679	−0.371 526	−0.147 592	.000 000	.079 492	.125 282	.154 156
6	13/2	.182 105	.079 980	.014 031	.000 000	.005 693	.015 890	.026 307

TABLE 3.4 (*contd.*)

N = 6 KAPPA = 0.0500 MU = 0.6200

K = 3/2, (611)

E		7.447 771	7.314 524	7.236 758	7.214 000	7.218 597	7.266 324	7.333 356
L	J							
2	3/2	−0.184 778	.218 894	.374 835	.000 000	−0.116 874	−0.134 882	−0.148 188
2	5/2	−0.576 967	.684 547	.863 103	1.000 000	.918 067	.853 170	.810 286
4	7/2	.558 405	−0.535 650	−0.219 296	.000 000	−0.319 722	−0.386 349	−0.401 964
4	9/2	.471 634	−0.389 437	−0.253 952	.000 000	.196 052	.303 141	.368 326
6	11/2	−0.286 777	.198 925	.037 185	.000 000	−0.050 809	−0.104 011	−0.138 107
6	13/2	−0.128 336	.072 986	.024 216	.000 000	.015 908	.043 941	.071 861

K = 3/2, (602)

E		7.909 669	7.668 182	7.477 390	7.464 000	7.605 962	7.780 663	7.965 431
L	J							
2	3/2	−0.466 853	−0.561 472	.783 387	1.000 000	.960 524	.925 833	.902 908
2	5/2	.536 559	.527 926	−0.423 336	.000 000	.187 120	.250 192	.281 778
4	7/2	.584 400	.569 126	−0.438 351	.000 000	.202 454	.271 644	.304 280
4	9/2	−0.323 705	−0.240 252	.104 386	.000 000	.031 493	.067 477	.094 825
6	11/2	−0.206 021	−0.151 198	.062 946	.000 000	.019 875	.042 732	.059 582
6	13/2	.073 485	.039 460	−0.009 032	.000 000	.002 215	.008 216	.015 193

K = 1/2, (660)

E		6.081 493	6.061 787	6.000 522	5.898 000	5.754 449	5.570 195	5.345 969
A		6.496 021	6.782 839	6.946 653	7.000 000	6.946 713	6.783 241	6.497 499
L	J							
0	1/2	.059 564	.017 964	.002 218	.000 000	−0.001 974	−0.014 180	−0.041 441
2	3/2	−0.002 906	.001 279	.000 421	.000 000	−0.000 834	−0.007 609	−0.026 798
2	5/2	.163 586	.077 166	.019 939	.000 000	.019 619	.074 411	.153 451
4	7/2	.005 591	.005 059	.001 664	.000 000	.002 380	.011 664	.031 886
4	9/2	.432 362	.308 099	.160 055	.000 000	−0.159 364	−0.305 379	−0.426 559
6	11/2	.010 214	.009 842	.005 953	.000 000	−0.008 142	−0.019 364	−0.035 579
6	13/2	.884 653	.947 984	.986 885	1.000 000	.986 986	.948 914	.888 697

K = 1/2, (651)

E		6.645 912	6.679 807	6.642 734	6.548 000	6.398 120	6.191 721	5.929 117
A		−3.750 801	−5.456 372	−5.894 265	−6.000 000	−5.875 625	−5.275 149	−4.078 335
L	J							
0	1/2	.276 866	−0.077 093	−0.008 305	.000 000	.007 508	.055 310	.135 007
2	3/2	−0.295 957	.147 176	.038 012	.000 000	.036 985	.138 485	.258 988
2	5/2	.222 818	−0.066 129	−0.010 918	.000 000	−0.018 552	−0.100 122	−0.205 100
4	7/2	−0.495 823	.398 326	.212 981	.000 000	−0.211 634	−0.392 252	−0.501 963
4	9/2	.145 882	−0.070 355	−0.026 350	.000 000	.045 123	.121 256	.153 916
6	11/2	−0.710 209	.896 706	.975 863	1.000 000	.975 265	.891 076	.753 353
6	13/2	−0.120 781	.018 076	−0.001 749	.000 000	.016 257	.071 816	.171 567

K = 1/2, (640)

E		6.775 600	6.798 651	6.771 195	6.680 000	6.530 147	6.331 527	6.112 957
A		2.163 504	4.210 156	4.847 362	5.000 000	4.827 707	4.061 249	2.702 230
L	J							
0	1/2	.400 514	.260 306	.067 878	.000 000	.053 904	.160 404	.234 334
2	3/2	−0.027 170	−0.029 254	.004 868	.000 000	.025 067	.086 257	.112 571
2	5/2	.547 689	.514 002	.295 480	.000 000	−0.287 589	−0.481 786	−0.546 507
4	7/2	.170 258	.033 735	.018 188	.000 000	−0.020 955	−0.007 032	.075 584

TABLE 3.4 (*contd.*)

N = 6 KAPPA = 0.0500 MU = 0.6200

L	J							
4	9/2	.478 768	.753 024	.939 125	1.000 000	.940 879	.779 686	.585 564
6	11/2	.381 316	.115 086	.024 798	.000 000	−0.057 539	−0.209 868	−0.383 401
6	13/2	−0.367 802	−0.292 845	−0.158 613	.000 000	.157 341	.287 591	.371 686

K = 1/2, (631)

E		7.119 125	7.129 836	7.185 541	7.130 000	6.957 304	6.739 765	6.558 203
A		−0.296 087	−0.831 847	−2.625 448	−4.000 000	−2.943 395	−1.931 183	−1.478 017
L	J							
0	1/2	−0.480 297	.553 898	−0.327 457	.000 000	−0.185 926	−0.292 080	−0.285 660
2	3/2	.509 866	−0.519 620	.425 148	.000 000	−0.410 576	−0.546 210	−0.565 077
2	5/2	.056 989	.179 816	−0.259 635	.000 000	.259 333	.208 242	.067 049
4	7/2	.255 288	−0.424 857	.774 887	1.000 000	.821 706	.593 308	.448 834
4	9/2	.464 137	−0.330 429	.091 578	.000 000	.128 600	.317 314	.415 695
6	11/2	−0.430 704	.307 314	−0.188 914	.000 000	.194 055	.338 233	.441 592
6	13/2	−0.200 006	.082 035	−0.009 219	.000 000	.014 510	.076 652	.159 165

K = 1/2, (620)

E		7.492 619	7.361 236	7.275 889	7.214 000	7.082 383	6.992 627	6.937 755
A		−0.364 787	−0.324 246	1.373 381	3.000 000	1.793 671	.879 459	.720 320
L	J							
0	1/2	.453 877	.460 335	.489 791	.000 000	−0.347 539	−0.405 373	−0.429 400
2	3/2	.320 914	.275 406	.080 576	.000 000	−0.052 897	.023 443	.048 887
2	5/2	.268 605	.413 070	.712 602	1.000 000	.812 045	.625 117	.514 125
4	7/2	.513 150	.566 013	.411 321	.000 000	−0.383 519	−0.505 089	−0.504 786
4	9/2	−0.459 905	−0.384 408	−0.261 172	.000 000	.252 349	.385 343	.461 101
6	11/2	−0.350 134	−0.258 403	−0.087 773	.000 000	−0.075 174	−0.188 782	−0.258 165
6	13/2	.146 397	.081 878	.026 661	.000 000	.024 169	.071 478	.121 773

K = 1/2, (611)

E		7.948 798	7.701 330	7.503 797	7.464 000	7.380 887	7.403 360	7.460 673
A		.008 745	−0.001 496	−0.140 076	−2.000 000	−1.102 324	−0.902 392	−0.795 053
L	J							
0	1/2	−0.266 588	−0.350 257	.543 282	.000 000	.389 641	.304 376	.251 825
2	3/2	.423 070	.466 334	−0.574 821	1.000 000	.773 007	.670 694	.618 105
2	5/2	.580 438	.588 524	−0.481 493	.000 000	.357 676	.460 955	.485 228
4	7/2	−0.518 060	−0.483 982	.357 692	.000 000	.339 073	.452 569	.491 495
4	9/2	−0.324 642	−0.245 727	.108 160	.000 000	.075 662	.170 450	.233 931
6	11/2	.184 658	.131 670	−0.053 534	.000 000	.044 466	.102 664	.143 472
6	13/2	.071 814	.039 180	−0.009 069	.000 000	.006 089	.025 166	.046 989

K = 1/2, (600)

E		8.470 452	8.101 353	7.754 322	7.500 000	7.630 710	7.804 805	7.989 326
A		−0.256 595	−0.379 034	−0.507 607	1.000 000	.353 254	.384 775	.431 356
L	J							
0	1/2	.501 624	.533 464	.594 172	1.000 000	.830 579	.792 923	.771 843
2	3/2	.608 051	.643 554	.693 447	.000 000	−0.478 637	−0.473 918	−0.464 520
2	5/2	−0.459 699	−0.419 706	−0.324 234	.000 000	.248 922	.325 124	.362 560
4	7/2	−0.358 134	−0.324 128	−0.238 173	.000 000	−0.132 323	−0.184 404	−0.208 507
4	9/2	.173 880	.125 477	.058 613	.000 000	.037 034	.076 812	.106 447
6	11/2	.088 874	.063 445	.027 974	.000 000	−0.013 977	−0.031 378	−0.044 282
6	13/2	−0.030 516	−0.016 523	−0.004 353	.000 000	.002 492	.008 899	.016 182

TABLE 3.4 (*contd.*)

N = 7 KAPPA = 0.0500 MU = 0.4340

DELTA =		−0.3	−0.2	−0.1	0.0	0.1	0.2	0.3
K = 15/2, (707)								
E		6.234 800	6.468 133	6.701 467	6.934 800	7.168 133	7.401 467	7.634 800
L	J							
7	15/2	1.000 000	1.000 000	1.000 000	1.000 000	1.000 000	1.000 000	1.000 000
K = 13/2, (716)								
E		6.503 626	6.650 516	6.793 863	6.934 800	7.074 057	7.212 112	7.349 286
L	J							
7	13/2	.147 679	.085 559	.037 525	.000 000	−0.029 774	−0.053 796	−0.073 491
7	15/2	.989 035	.996 333	.999 296	1.000 000	.999 557	.998 552	.997 296
K = 13/2, (707)								
E		7.015 974	7.235 751	7.459 070	7.684 800	7.912 210	8.140 821	8.370 314
L	J							
7	13/2	.989 035	.996 333	.999 296	1.000 000	.999 557	.998 552	.997 296
7	15/2	−0.147 679	−0.085 559	−0.037 525	.000 000	.029 774	.053 796	.073 491
K = 11/2, (725)								
E		6.669 673	6.785 303	6.868 951	6.934 800	6.990 722	7.040 644	7.086 662
L	J							
5	11/2	.461 459	.246 633	.094 907	.000 000	−0.060 760	−0.101 737	−0.130 755
7	13/2	.076 016	.070 194	.038 578	.000 000	−0.036 604	−0.068 732	−0.096 216
7	15/2	.883 899	.966 563	.994 738	1.000 000	.997 481	.992 434	.986 735
K = 11/2, (716)								
E		7.000 263	7.179 119	7.383 157	7.599 000	7.795 264	7.924 135	8.050 111
L	J							
5	11/2	.862 076	.954 102	.988 543	1.000 000	.579 826	.255 058	.192 058
7	13/2	−0.273 676	−0.192 429	−0.121 468	.000 000	.812 134	,962 462	.973 967
7	15/2	−0.426 529	−0.229 479	−0.089 605	.000 000	.065 121	.092 803	.120 421
K = 11/2, (705)								
E		7.348 665	7.454 178	7.566 492	7.684 800	7.832 614	8.053 821	8.281 827
L	J							
5	11/2	.209 479	.169 887	.117 372	.000 000	.812 472	.961 559	.972 634
7	13/2	.958 813	.978 797	.991 845	1.000 000	−0.582 322	−0.262 570	−0.205 256
7	15/2	−0.191 822	−0.114 432	−0.049 664	.000 000	.028 121	.080 387	.108 872
K = 9/2, (734)								
E		6.826 180	6.895 168	6.929 612	6.934 800	6.918 972	6.889 077	6.849 809
L	J							
5	9/2	.091 475	.034 810	.006 995	.000 000	.004 153	.012 868	.022 795
5	11/2	.451 639	.297 634	.136 376	.000 000	−0.102 094	−0.175 456	−0.228 259
7	13/2	.079 331	.060 917	.034 532	.000 000	−0.037 423	−0.073 252	−0.105 399
7	15/2	.883 946	.952 098	.990 030	1.000 000	.994 062	.981 674	.967 610

TABLE 3.4 (*contd.*)

N = 7 KAPPA = 0.0500 MU = 0.4340

K = 9/2, (725)

L	J							
	E	7.363 397	7.427 278	7.504 759	7.599 000	7.687 376	7.724 402	7.751 051
5	9/2	.519 972	.149 451	.065 211	.000 000	−0.095 522	−0.156 853	−0.186 308
5	11/2	.736 412	.925 358	.978 703	1.000 000	.785 071	.391 446	.278 300
7	13/2	.023 478	−0.205 107	−0.144 666	.000 000	.603 141	.896 036	.926 610
7	15/2	−0.432 175	−0.281 616	−0.130 231	.000 000	.103 735	.138 881	.170 973

K = 9/2, (714)

L	J							
	E	7.397 252	7.551 467	7.642 511	7.684 800	7.732 549	7.837 682	7.958 715
5	9/2	.580 233	.539 101	.199 008	.000 000	−0.035 535	−0.027 818	−0.060 777
5	11/2	−0.388 601	.068 999	.123 286	.000 000	−0.608 537	.898 843	.925 789
7	13/2	.711 865	.834 057	.970 807	1.000 000	.792 062	−0.417 668	−0.324 316
7	15/2	.074 617	−0.094 645	−0.052 250	.000 000	−0.032 532	.129 850	.184 499

K = 9/2, (705)

L	J							
	E	7.780 772	7.827 021	7.957 385	8.149 000	8.362 036	8.583 105	8.808 025
5	9/2	−0.620 154	.828 144	.977 801	1.000 000	.994 784	.987 146	.980 345
5	11/2	.320 482	−0.224 422	−0.091 339	.000 000	.054 073	.089 816	.115 592
7	13/2	.697 427	−0.508 496	−0.188 184	.000 000	.086 366	.131 561	.158 441
7	15/2	−0.162 161	.072 413	.012 237	.000 000	.004 649	.012 931	.021 431

K = 7/2, (743)

L	J							
	E	6.935 332	6.979 411	6.977 086	6.934 800	6.859 737	6.759 946	6.642 558
3	7/2	.206 811	.076 941	.015 234	.000 000	.008 700	.026 004	.044 256
5	9/2	.044 893	.027 440	.007 555	.000 000	.006 482	.022 040	.041 339
5	11/2	.484 596	.335 292	.164 543	.000 000	−0.138 162	−0.243 594	−0.320 374
7	13/2	.051 564	.046 955	.028 178	.000 000	−0.033 979	−0.069 586	−0.103 435
7	15/2	.847 184	.937 391	.985 821	1.000 000	.989 767	.966 777	.939 678

K = 7/2, (734)

L	J							
	E	7.367 420	7.511 261	7.584 107	7.599 000	7.592 102	7.543 697	7.475 690
3	7/2	.786 385	.628 424	.256 561	.000 000	−0.118 917	−0.119 293	−0.108 758
5	9/2	−0.208 024	−0.074 335	.026 988	.000 000	−0.123 903	−0.256 219	−0.318 288
5	11/2	.368 311	.685 937	.942 030	1.000 000	.875 418	.509 533	.339 606
7	13/2	−0.246 649	−0.220 344	−0.145 961	.000 000	.429 975	.789.158	.848 188
7	15/2	−0.376 610	−0.283 718	−0.157 233	.000 000	.138 819	.194 236	.228 274

K = 7/2, (723)

L	J							
	E	7.610 110	7.670 722	7.701 528	7.684 800	7.648 293	7.644 212	7.660 791
3	7/2	.157 725	.029 506	.008 256	.000 000	.054 127	−0.164 352	−0.222 010
5	9/2	.594 861	.481 284	.238 469	.000 000	−0.104 887	.038 419	−0.013 859
5	11/2	.227 984	.233 817	.131 229	.000 000	−0.439 693	.792 012	.836 285
7	13/2	.714 005	.832 291	.960 838	1.000 000	.889 833	−0.563 749	−0.435 326
7	15/2	−0.243 892	−0.141 834	−0.051 323	.000 000	−0.030 617	.162 527	.248 271

TABLE 3.4 (*contd.*)

N = 7 KAPPA = 0.0500 MU = 0.4340

K = 7/2, (714)

L	J							
E		7.815 385	7.814 257	7.911 805	8.089 600	8.216 547	8.317 503	8.429 040
3	7/2	−0.497 891	.729 072	.938 850	1.000 000	.427 260	.301 199	.250 124
5	9/2	.256 177	−0.254 239	−0.239 074	.000 000	.886 377	.910 629	.903 426
5	11/2	.676 053	−0.547 058	−0.231 858	.000 000	.115 059	.171 046	.209 136
7	13/2	−0.406 081	.296 739	.084 197	.000 000	.135 728	.223 289	.273 915
7	15/2	−0.254 024	.128 411	.023 616	.000 000	.011 160	.030 308	.049 930

K = 7/2, (703)

L	J							
E		8.228 953	8.148 216	8.116 007	8.149 000	8.307 188	8.525 175	8.749 122
3	7/2	.257 011	.258 354	.228 993	.000 000	.894 596	.931 322	.935 074
5	9/2	.731 590	.835 134	.940 846	1.000 000	−0.433 522	−0.321 161	−0.283 925
5	11/2	−0.347 114	−0.251 250	−0.120 522	.000 000	.089 363	.156 518	.197 275
7	13/2	−0.511 664	−0.410 465	−0.218 181	.000 000	−0.061 176	−0.068 674	−0.073 079
7	15/2	.128 187	.064 777	.015 603	.000 000	.005 350	.016 765	.027 666

K = 5/2, (752)

L	J							
E		7.021 437	7.041 677	7.012.156	6.934 800	6.814 010	6.656 473	6.470 257
3	5/2	.059 605	.016 299	.001 746	.000 000	−0.001 064	−0.006 226	−0.015 096
3	7/2	.207 262	.096 403	.023 199	.000 000	.017 463	.056 384	.100 264
5	9/2	.046 010	.022 704	.006 445	.000 000	.007 128	.026 992	.054 586
5	11/2	.494 735	.357 383	.183 537	.000 000	−0.167 764	−0.303 046	−0.402 219
7	13/2	.041 591	.034 019	.020 699	.000 000	−0.027 101	−0.058 293	−0.090 328
7	15/2	.839 571	.927 925	.982 498	1.000 000	.985 274	.949 115	.903 770

K = 5/2, (743)

L	J							
E		7.583 060	7.638 483	7.643 570	7.599 000	7.513 264	7.385 489	7.227 713
3	5/2	.536 719	.153 314	.033 721	.000 000	.024 569	.072 213	.104 943
3	7/2	.524 034	.545 108	.299 748	.000 000	−0.211 772	−0.256 223	−0.232 468
5	9/2	.302 231	.005 326	.014 974	.000 000	−0.121 441	−0.312 812	−0.419 594
5	11/2	.379 220	.736 372	.928 438	1.000 000	.909 354	.600 422	.372 889
7	13/2	.170 737	−0.152 270	−0.122 791	.000 000	.291 235	.636 821	.731 181
7	15/2	−0.415 955	−0.337 481	−0.178 088	.000 000	.167 506	.255 415	.291 916

K = 5/2, (732)

L	J							
E		7.636 550	7.727 421	7.743 442	7.684 800	7.580 787	7.477 994	7.392 695
3	5/2	−0.266 475	.280 241	.054 567	.000 000	.014 211	−0.001 466	.025 852
3	7/2	.366 535	.036 642	.022 985	.000 000	.061 155	0.238 298	−0.353 505
5	9/2	−0.499 631	.495 649	.263 412	.000 000	−0.176 802	.159 122	.096 835
5	11/2	.326 721	.048 094	.104 744	.000 000	−0.304 005	.641 077	.701 935
7	13/2	−0.629 300	.816 906	.956 225	1.000 000	.933 654	−0.690 921	−0.535 359
7	15/2	−0.205 539	−0.069 328	−0.042 080	.000 000	−0.025 872	.171 878	.292 688

K = 5/2, (723)

E		7.985 616	8.027 311	8.081 789	8.089 600	8.084 761	8.067 018	8.066 243

TABLE 3.4 (*contd.*)

N = 7 KAPPA = 0.0500 MU = 0.4340

L	J							
3	5/2	−0.692 847	.802 511	.219 428	.000 000	−0.224 621	−0.305 966	−0.340 531
3	7/2	.089 291	.160 454	.915 224	1.000 000	.589 477	.392 530	.296 400
5	9/2	.013 605	.237 331	−0.166 879	.000 000	.735 026	.768 219	.742 781
5	11/2	.352 144	−0.323 075	−0.288 720	.000 000	.185 038	.262 960	.307 992
7	13/2	.586 092	−0.398 781	.044 488	.000 000	.164 801	.299 913	.375 879
7	15/2	−0.210 135	.102 476	.032 092	.000 000	.020 032	.055 208	.091 205

K = 5/2, (712)

	E	8.264 784	8.152 327	8.140 579	8.149 000	8.173 600	8.265 632	8.374 395
L	**J**							
3	5/2	−0.039 685	.050 331	.586 474	.000 000	.031 576	−0.024 043	−0.058 409
3	7/2	−0.615 386	−0.725 422	−0.007 870	.000 000	.768 290	.832 087	.827 038
5	9/2	.426 935	.448 296	.771 500	1.000 000	−0.612 867	−0.467 225	−0.418 840
5	11/2	.537 658	.422 822	−0.060 764	.000 000	.133 738	.256 012	.327 665
7	13/2	−0.346 471	−0.289 150	−0.238 687	.000 000	−0.123 031	−0.147 951	−0.161 005
7	15/2	−0.168 324	−0.088 734	.010 462	.000 000	.010 238	.036 354	.062 305

K = 5/2, (703)

	E	8.705 353	8.509 582	8.375 263	8.439 600	8.630 377	8.844 194	9.065 496
L	**J**							
3	5/2	−0.394 487	−0.501 136	.777 034	1.000 000	.973 520	.948 974	.932 050
3	7/2	.401 758	.374 484	−0.267 186	.000 000	.115 562	.167 140	.197 723
5	9/2	.688 803	.704 622	−0.554 335	.000 000	.195 125	.260 077	.290 575
5	11/2	−0.302 656	−0.211 102	.079 335	.000 000	.019 660	.044 582	.065 092
7	13/2	−0.330 990	−0.256 296	.105 720	.000 000	.021 006	.043 040	.058 301
7	15/2	.085 821	.043 358	−0.008 483	.000 000	.001 517	.005 777	.010 878

K = 3/2, (761)

	E	7.075 024	7.082 460	7.035 301	6.934 800	6.782 781	6.582 749	6.340 811
L	**J**							
1	3/2	.099 643	.028 479	.003 235	.000 000	−0.002 248	−0.013 796	−0.034 208
3	5/2	.011 443	.007 168	.001 260	.000 000	−0.001 470	−0.010 684	−0.030 168
3	7/2	.231 584	.110 987	.028 520	.000 000	.025 676	.089 594	.167 784
5	9/2	.019 341	.013 402	.004 234	.000 000	.005 632	.024 128	.054 452
5	11/2	.503 691	.371 086	.195 383	.000 000	−0.189 088	−0.348 804	−0.466 078
7	13/2	.021 313	.020 169	.012 613	.000 000	−0.017 500	−0.039 640	−0.065 214
7	15/2	.825 697	.921 156	.980 216	1.000 000	.981 449	.931 585	.863 320

K = 3/2, (752)

	E	7.585 464	7.684 610	7.681 229	7.599 000	7.455 213	7.256 805	7.013 559
L	**J**							
1	3/2	.593 650	.344 307	.077 394	.000 000	.042 965	.105 073	.130 796
3	5/2	−0.210 712	−0.042 742	.012 842	.000 000	.034 598	.136 839	.228 433
3	7/2	.515 283	.553 826	.330 789	.000 000	−0.289 730	−0.406 053	−0.370 220
5	9/2	−0.247 965	−0.090 556	.002 635	.000 000	−0.091 909	−0.299 220	−0.456 411
5	11/2	.277 482	.649 599	.916 765	1.000 000	.916 276	.647 315	.365 903
7	13/2	−0.249 156	−0.177 435	−0.085 264	.000 000	.172 371	.447 087	.570 739
7	15/2	−0.370 272	−0.333 528	−0.191 546	.000 000	.187 862	.311 319	.354 555

TABLE 3.4 (*contd.*)

N = 7 KAPPA = 0.0500 MU = 0.4340

K = 3/2, (741)

E		7.742 324	7.781 365	7.770 820	7.684 800	7.532 219	7.345 233	7.161 378
L	J							
1	3/2	.088 025	−0.034 006	−0.006 042	.000 000	−0.003 373	−0.046 626	.111 713
3	5/2	.335 544	.230 263	.062 745	.000 000	.038 804	060 111	−0.021 657
3	7/2	.277 030	.099 468	.018 182	.000 000	.047 233	.242 566	.415 800
5	9/2	.539 460	.488 319	276 767	.000 000	−0.238 883	−0.310 571	.261 688
5	11/2	.236 556	.180 522	.075 057	.000 000	−0.185 432	−0.459 271	.529 783
7	13/2	.620 661	.807 918	.955 311	1.000 000	.951 029	.777 317	−0.608 382
7	15/2	−0.265 932	−0.110 242	−0.029 038	.000 000	−0.018 582	−0.154 171	.308 030

K = 3/2, (732)

E		8.000 013	8.018 300	8.106 611	8.089 600	7.972 328	7.833 497	7.718 974
L	H							
1	3/2	−0.581 069	.725 300	.610 579	.000 000	−0.233 689	−0.250 282	−0.236 994
3	5/2	.337 653	−0.333 823	−0.167 158	.000 000	−0.322 940	−0.474 296	−0.521 505
3	7/2	.125 696	.188 158	.670 614	1.000 000	.667 464	.381 788	.223 160
5	9/2	.000 155	−0.152 517	−0.252 133	.000 000	.550 726	.557 367	.488 085
5	11/2	.501 138	−0.437 152	−0.274 425	.000 000	.251 866	.353 509	.396 982
7	13/2	−0.460 332	.309 975	.097 694	.000 000	.167 223	.351 344	.452 455
7	15/2	−0.263 637	.129 041	.033 220	.000 000	.029 866	.087 012	.146 726

K = 3/2, (721)

E		8.364 178	8.266 880	8.209 633	8.149 000	8.058 404	8.026 361	8.021 191
L	J							
1	3/2	.315 311	.247 057	.045 365	.000 000	.157 438	−0.283 087	−0.341 930
3	5/2	.562 558	.573 005	.488 910	.000 000	−0.182 832	.127 678	.089 760
3	7/2	.142 741	.289 238	.335 058	.000 000	−0.610 599	.665 914	.619 971
5	9/2	.353 764	.495 998	.750 955	1.000 000	.714 047	−0.538 032	−0.476 020
5	11/2	−0.411 128	−0.334 453	−0.152 235	.000 000	−0.154 219	.327 909	.422 217
7	13/2	−0.490 811	−0.401 387	−0.243 204	.000 000	.187 350	−0.244 069	−0.272 890
8	15/2	.169 296	.089 360	.019 703	.000 000	−0.014 408	.059 554	.106 450

K = 3/2, (712)

E		8.733 652	8.517 049	8.360 467	8.406 600	8.433 284	8.510 263	8.611 457
L	J							
1	3/2	−0.307 227	−0.425 063	.699 550	1.000 000	.438 449	.345 337	.289 797
3	5/2	.239 839	.237 662	−0.298 525	.000 000	.800 319	.750 832	.717 103
3	7/2	.617 021	.638 365	−0.509 030	.000 000	.247 375	.330 191	.364 609
5	9/2	−0.493 972	−0.479 639	.375 880	.000 000	.318 450	.429 770	.472 363
5	11/2	−0.383 903	−0.288 823	.117 518	.000 000	.049 772	.112 676	.160 191
7	13/2	.256 940	.197 501	−0.084 748	.000 000	.046 393	.101 338	.137 764
7	15/2	.099 822	.053 383	−0.011 072	.000 000	.004 321	.017 339	.032 775

K = 3/2, (701)

E		9.202 745	8.886 069	8.606 006	8.439 600	8.602 503	8.815 158	9.036 030

TABLE 3.4 (*contd.*)

N = 7 KAPPA = 0.0500 MU = 0.4340

L	J							
1	3/2	.313 769	.334 231	.360 177	.000 000	.852 352	.851 210	.843 907
3	5/2	.595 715	.669 959	.799 878	1.000 000	−0.468 047	−0.415 381	−0.390 099
3	7/2	−0.432 685	−0.380 122	−0.261 412	.000 000	.183 392	.264 624	.307 883
5	9/2	−0.527 223	−0.503 821	−0.393 173	.000 000	−0.141 007	−0.169 094	−0.179 707
5	11/2	.213 992	146 116	.058 647	.000 000	.024 518	.056 567	.081 813
7	13/2	.182 830	.138 476	.063 853	.000 000	−0.017 594	−0.032 229	−0.041 380
7	15/2	−0.047 674	−0.024 312	−0.005 424	.000 000	.001 673	.006 580	.012 347

K = 1/2, (770)

E	7.102 821	7.102 758	7.046 806	6.934 800	6.766 906	6.543 665	6.266 655
A	−7.190 609	−7.652 545	−7.914 838	−8.000 000	−7.915 213	−7.656 586	−7.214 087

L	J							
1	1/2	.038 101	.008 071	.000 511	.000 000	.000 443	.006 044	.024 398
1	3/2	.094 629	.031 621	.004 209	.000 000	−0.004 039	−0.029 006	−0.082 433
3	5/2	.017 570	.004 260	.000 513	.000 000	−0.000 775	−0.007 644	−0.028 842
3	7/2	.235 630	.117 110	.031 164	.000 000	.030 793	.114 098	.225 709
5	9/2	.012 058	.005 042	.001 478	.000 000	.002 167	.010 655	.029 513
5	11/2	.506 909	.377 340	.201 082	.000 000	.200 345	−0.374 650	−0.502 463
7	13/2	.009 351	.006 875	.004 235	.000 000	−0.006 058	−0.014 422	−0.026 576
7	15/2	.822 541	.918 011	.979 059	1.000 000	.979 211	.919 435	.828 726

K = 1/2, (761)

E	7.656 279	7.724 840	7.700 184	7.599 000	7.423 385	7.173 695	6.849 178
A	−1.622 006	−5.108 859	−5.800 582	−6.000 000	−5.758 468	−4.196 714	−0.895 636

L	J							
1	1/2	.405 555	.140 684	.017 716	.000 000	−0.014 815	−0.094 578	−0.198 224
1	3/2	.402 590	.286 085	.085 557	.000 000	.080 146	.246 409	.335 461
3	5/2	.304 790	.067 450	.007 778	.000 000	.021 014	.138 061	.314 741
3	7/2	.471 521	.558 255	.343 177	.000 000	−0.337 876	−0.525 910	−0.469 300
4	9/2	.278 678	.038 653	.001 477	.000 000	−0.036 651	−0.183 773	−0.380 160
5	11/2	.276 469	.671 940	.913 340	1.000 000	.913 534	.656 107	.309 560
7	13/2	.238 270	.013 838	−0.028 886	.000 000	.059 424	.222 484	.368 278
7	15/2	−0.383 861	−0.359 131	−0.198 766	.000 000	.198 335	.347 775	.391 012

K = 1/2, (750)

E	7.750 645	7.798 729	7.784 155	7.684 800	7.505 966	7.256 121	6.977 997
A	2.509 059	6.101 290	6.810 046	7.000 000	6.765 949	5.185 258	1.799 659

L	J							
1	1/2	−0.095 912	.126 068	.016 994	.000 000	−0.011 511	−0.042 923	−0.010 504
1	3/2	.190 523	.010 510	.002 204	.000 000	.000 970	−0.037 723	−0.182 629
3	5/2	−0.300 098	.245 646	.068 083	.000 000	.063 633	.180 989	.185 477
3	7/2	.341 437	−0.033 050	.007 507	.000 000	.017 736	.158 230	.380 714
5	9/2	−0.501 652	.495 673	.283 088	.000 000	−0.278 039	−0.451 100	−0.442 186
5	11/2	.269 371	−0.063 418	.023 728	.000 000	−0.066 848	−0.250 782	−0.343 654
7	13/2	−0.596 371	.820 058	.956 148	1.000 000	.955 861	.813 572	.625 155
7	15/2	−0.260 749	.018 810	−0.009 729	.000 000	−0.007 646	−0.103 238	−0.287 656

K = 1/2, (741)

E	8.058 472	8.091 459	8.162 036	8.089 600	7.892 831	7.619 234	7.381 248
A	.863 103	.495 490	−2.968 038	−4.000 000	−1.816 260	.425 086	.863 310

TABLE 3.4 (*contd.*)

N = 7 KAPPA = 0.0500 MU = 0.4340

L	J							
1	1/2	−0.578 544	.606.237	.283 989	.000 000	.190 275	.344 050	.375 931
1	3/2	−0.213 263	.388 734	.506 118	.000 000	−0.439 616	−0.435 054	−0.347 295
3	5/2	−0.329 830	.419 407	.164 870	.000 000	−0.297 618	−0.487 809	−0.488 337
3	7/2	.158 960	.122 061	.720 315	1.000 000	.681 673	.245 403	.002 927
5	9/2	.072 555	.146 862	.110 268	.000 000	.333 140	.289 191	.142 546
5	11/2	.403 215	−0.373 741	−0.317 962	.000 000	.297 555	.408 550	.440 322
7	13/2	.509 016	−0.339 296	−0.047 958	.000 000	.127 913	.360 061	.489 391
7	15/2	−0.242 499	.119 119	.040 007	.000 000	.037 362	.118 276	.214 182

K = 1/2 (730)

E	8.375 594	8.268 003	8.213 936	8.149 000	7.967 815	7.810 396	7.691 026
A	.015 526	.322 794	3.823 577	5.000 000	2.739 386	.440 631	−0.092 514

L	J							
1	1/2	−0.040 434	−0.037 094	.226 224	.000 000	.075 476	.008 567	.040 477
1	3/2	−0.495 750	.461 708	−0.093 584	.000 000	.174 118	.400 574	−0.466 822
3	5/2	.410 507	−0.422 959	.493 123	.000 000	−0.385 608	−0.359 240	.336 551
3	7/2	−0.199 374	.376 916	−0.234 587	.000 000	−0.423 344	−0.451 894	.328 556
5	9/2	.307 072	−0.415 391	.753 123	1.000 000	.746 275	.499 154	−0.405 008
5	11/2	.476 022	−0.390 446	.077 005	.000 000	−0.144 042	−0.366 656	.462 730
7	13/2	−0.434 704	.360 298	−0.262 050	.000 000	.241 132	.339 893	−0.395 544
7	15/2	−0.185 667	.098 375	−0.008 326	.000 000	−0.015 939	−0.084 185	.156 898

K = 1/2, (721)

E	8.779 293	8.570 763	8.428 215	8.406 600	8.239 939	8.175 546	8.159 058
A	.776 771	.975 325	.880 562	−2.000 000	.749 384	.927 589	.900 031

L	J							
1	1/2	−0.507 385	−0.569 567	.700 143	.000 000	−0.505 553	−0.507 176	−0.494 445
1	3/2	.152 395	.026 339	.266 457	1.000 000	.444 063	.223 865	.107 287
3	5/2	.070 560	−0.001 003	.251 474	.000 000	.473 923	.323 800	.257 574
3	7/2	.418 043	.467 484	−0.397 277	.000 000	.382 377	.472 547	.468 724
5	9/2	.531 822	.558 611	−0.442 110	.000 000	.402 054	.530 077	.548 968
5	11/2	−0.370 414	−0.276 677	.110 003	.000 000	.092 465	.217 753	.298 485
7	13/2	−0.328 166	−0.253 985	.100 205	.000 000	.078 303	.185 786	.248 480
7	15/2	.108 918	.057 028	−0.011 356	.000 000	.008 924	.039 948	.075 925

K = 1/2, (710)

E	9.218 557	8.892 054	8.601 142	8.439 600	8.418 040	8.492 805	8.593 203
A	−0.022 975	.008 329	.102 756	3.000 000	.227 519	−0.042 872	−0.194 546

L	J							
1	1/2	−0.048 216	−0.017 617	.045 280	.000 000	−0.029 603	.022 989	.054 515
1	3/2	−0.480 461	−0.558 052	−0.668 783	.000 000	−0.681 276	−0.627 820	−0.581 733
3	5/2	.467 155	.492 484	.560 579	1.000 000	.603 864	.532 427	.500 509
3	7/2	.512 909	.475 737	.346 788	.000 000	−0.291 291	−0.414 897	−0.466 500
5	9/2	−0.448 583	−0.419 380	−0.328 648	.000 000	.284 068	.353 998	.377 263
5	11/2	−0.235 928	−0.167 527	−0.069 644	.000 000	−0.052 889	−0.126 828	−0.182 270
7	13/2	.161 429	.121 433	.057 191	.000 000	.044 436	.089 199	.117 029
7	15/2	.050 734	.026 656	.006 072	.000 000	−0.004 333	−0.018 427	−0.035 190

TABLE 3.4 (*contd.*)

N = 7 KAPPA = 0.0500 MU = 0.4340

K = 1/2, (701)

E		9.718 339	9.278 061	8.856 859	8.556 600	8.711 785	8.921 871	9.141 636
A		.671 131	.858 177	1.066 518	1.000 000	1.007 693	.917 608	.833 783
L	J							
1	1/2	−0.478 290	−0.520 223	.612 634	1.000 000	.837 428	.782 928	.754 713
1	3/2	.497 344	.490 551	−0.457 683	.000 000	.329 624	.378 172	.398 568
3	5/2	.562 333	.581 165	−0.589 460	.000 000	.411 077	.439 076	.443 972
3	7/2	−0.320 488	−0.267 793	.170 956	.000 000	.098 063	.159 664	.196 440
5	9/2	−0.294 820	−0.265 980	.192 439	.000 000	.105 335	.153 429	.176 163
5	11/2	.109 543	.073 099	−0.028 848	.000 000	.014 673	.037 664	.057 336
7	13/2	.075 266	.055 456	−0.025 439	.000 000	.012 239	.027 378	.038 068
7	15/2	−0.019 307	−0.009 859	.002 260	.000 000	.001 067	.004 619	.009 087

4

NUCLEAR CORRELATIONS

4.1. Introduction

So far we have introduced two extreme models of nuclear structure – the completely collective liquid-drop model and the extreme individual-particle model. The success of the spherical shell model has been its association of the shell closures with the observed magic-number nuclei and the predictions of spins and moments of the closed shell plus and minus one nucleon nuclei. The extreme collective model is successful in explaining a large range of ground-state energy systematics, qualitatively it explains the observed rotational and vibrational spectra, especially in the heavy nuclei, besides providing us qualitatively with a mechanism for describing fission.

The deformed shell model is again an independent-particle model, but as we indicated at the end of the last chapter it is possible to extract apparently collective features (e.g. the rotational spectra) from it. Thus we can begin to build a link between these two extreme models. It is this microscopic description of collective phenomena, implying as it does strong correlations in the nucleon motions, that we shall seek to develop in the present chapter. We shall see that in order to extract the essential collectivity of nuclear behaviour we shall have to study highly simplified versions of the microscopic models. However, these studies will provide us with a qualitative understanding of some nuclear systematics, while a quantitative study would require a major numerical calculation of the more exact theory.

4.2. Hartree–Fock theory: static and time-dependent (Brown 1967; Irvine 1972)

The dominant collectivity of the nucleus is not its vibrational or rotational modes of excitation but that it exists at all as a stable or metastable entity. According to Bethe (Bethe 1971) there are three dominant aspects of the nucleon–nucleon interaction which ensure nuclear saturation, i.e. the leading term in the mass formula goes as A and not A^2 (see section 3.2.1), which is equivalent to the observation that nuclei are incompressible and that the central densities are essentially independent of mass number. The three factors involved are, in decreasing order of importance: (1) the tensor force, (2) the exchange character of the force, and (3) the repulsive core.

To illustrate the basic problem let us consider an infinitely large nucleus, so that we may neglect surface effects. We shall assume that the number of neutrons

N equals the number of protons Z and that there is no Coulomb interaction between the protons – such a model is known as nuclear matter. The single-particle states in such a translationally invariant system are momentum eigenstates, and the density ρ is given by the Fermi momentum k_F,

$$\rho = \left(\frac{2}{3\pi^2}\right) k_F{}^3. \tag{4.1}$$

The kinetic energy per particle is proportional to $k_F{}^2$, and the potential energy per particle is proportional to the number of particles within the interaction volume of any one nucleon, which is in turn proportional to the density and hence to $k_F{}^3$. If the nucleon–nucleon interaction was a simple central attractive force, then the total energy per particle would be

$$E/A = ak_F{}^2 - bk_F{}^3, \tag{4.2}$$

where a and b are positive constants, and hence at large densities the $k_F{}^3$ term will begin to dominate and the binding energy would increase without limit as the density increases. Of course, if the Coulomb interaction is switched on this would yield a correction to (4.2) of the form $ck_F{}^6$, where $c \ll b$, which would be dominant at high densities, and hence saturation would be achieved. However, this cannot be the answer since there are many stable nuclei with Coulomb energies which are negligible compared to their nuclear binding energies. Thus we return to Bethe's three criteria for the saturation of nuclear forces.

1. *The tensor force.* This has the same form as the more familiar dipole–dipole interaction of electro- or magnetostatics, in which the spin of the nucleons takes the role of the dipole moment.

$$V_T = V(r)\left(\frac{\boldsymbol{\sigma}_1 \cdot \mathbf{r}_{12}\ \boldsymbol{\sigma}_2 \cdot \mathbf{r}_{12}}{r_{12}^2} - \tfrac{1}{3}\boldsymbol{\sigma}_1 \cdot \boldsymbol{\sigma}_2\right). \tag{4.3}$$

This acts between nucleons in triplet spin states. Thus between like particles, which must be in triplet isospin states, this acts only in odd relative partial waves, while between neutrons and protons, which can be in either isospin triplet or isospin singlet states, it can act in either even or odd partial waves. In scattering through the tensor force the relative orbital angular momentum $\mathbf{L}$ of the pair of particles need not be conserved, but of course the total angular momentum $\mathbf{J} = \mathbf{L} + \mathbf{S}$ is conserved. At low densities the mean separation of nucleons is large while the range of the nucleon–nucleon interaction is small. Hence the nucleons interact with each other through those relative partial waves which are of shortest range, i.e. relative S states. A principal contributor to the nuclear binding energy is the tensor force acting in the ${}^3S_1 - {}^3D_1$ channel, i.e. a pair of nucleons in a relative ${}^3S_1$ state scatter through the tensor force into a ${}^3D_1$ state.† In second-order perturbation theory they can then reinteract and

† This explains why the deuteron is bound while the diproton and dineutron are not, and also why the deuteron has a measurable quadrupole moment.

scatter back into the original $^3S_1$ state. Schematically this gives a contribution to the nuclear binding energy of the form

$$\Delta E_T^{(2)} = \sum_n |\langle \Phi_0 \mid V_T \mid \Phi_n \rangle|^2/(E_n - E_0). \tag{4.4}$$

The amplitude for scattering into an intermediate state Φ_n is essentially the Fourier transform of the potential. At nuclear densities $\Delta E_T{}^{(2)}$ can contribute as much as 50 per cent of the binding energy. As the density increases, i.e. the significant range decreases, the scattering amplitude peaks at higher and higher relative momentum. Hence the energy denominator $E_n - E_0$ increases sharply and the contribution form $\Delta E_T{}^{(2)}$ is damped.

2. *The exchange character.* The exchange mixture presented in eqn (3.52) implies that the nucleon–nucleon interaction is different in even and odd spatial states. Matrix elements of the interaction have coefficients of the form

$$\begin{aligned}&(W{+}M{+}B{-}H) \text{ space symmetric, spin triplet, isospin singlet,}\\&(W{+}M{-}B{+}H) \text{ space symmetric, spin singlet, isospin triplet,}\\&(W{-}M{+}B{+}H) \text{ space antisymmetric, spin triplet, isospin triplet,}\\&(W{-}M{-}B{-}H) \text{ space antisymmetric, spin singlet, isospin singlet.}\end{aligned} \tag{4.5}$$

More specifically, it is attractive in space symmetric states (even angular momentum partial waves) and repulsive in space antisymmetric states (odd partial waves). We have seen that at low densities and up to nuclear densities the dominant interaction is in relative S states (an attraction). As the density is increased relative P state interactions become important, and these are repulsions.

3. *The repulsive core.* The most successful potential parameterizations of the nucleon–nucleon interaction involve strong repulsive cores at short distances $\sim 0{\cdot}4$ fm. Clearly, these cores set a limit on the density of the nucleus. However, the fact that the core radius is only 0·4 fm while the mean separation of nearest nucleons at nuclear densities is $\sim 1{\cdot}2$ fm indicates that under normal circumstances the repulsive core cannot be dominant in ensuring saturation. However, it may play a role in the stability of neutron stars.

We have discussed in section 3.3.1 how, if the nucleus saturates, we may obtain a zero-order approximation to the nuclear eigenstates as Slater-determinantal wavefunctions of individual nucleons moving in their mutual mean field – the static Hartree-Fock approximation. First-order approximate wavefunctions may then be obtained by allowing the zero-order wavefunctions to be mixed via the residual interaction. The question arises as to which configurations are likely to be of the most importance.

In heavy nuclei we are interested in excitations of at most a few MeV. Typical matrix elements of the residual interaction are less than 1 MeV, and the Fermi energy of nucleons at nuclear densities are of the order tens of MeV. Thus, just as relatively few electrons contribute to the conduction properties of metals in low field at room temperature, we would expect relatively few nucleons near the Fermi surface to contribute to the excitations of a heavy nucleus. Our first guess, therefore, might be that the nucleons inside the ^{208}Pb closed-shell configuration would be inert and that excitations would arise through the rearrangements of the valence nucleons outside this core.

We have seen that in the static Hartree–Fock theory we generally will have a non-spherical equilibrium density ρ. If the nucleus is behaving collectively this density will be a function of time, and the Heisenberg equation of motion is (Green 1965)

$$i\hbar\, \partial\rho/\partial t = [\mathcal{H}, \rho(t)]\,. \tag{4.6}$$

If we now approximate $\mathcal{H}$ by the Hartree–Fock Hamiltonian

$$\mathcal{H} \rightarrow \mathcal{H}^{\mathrm{HF}} = \sum_{j} \epsilon_{ij}\, \rho_{ij}, \tag{4.7}$$

where ρ_{ij} is the one-body density matrix and

$$\epsilon_{ij} = \langle i \,|-\frac{\hbar^2}{2m}\nabla^2\; | j\rangle + \sum_{k} \langle ik \,|\, V\big(\, | jk\rangle - | kj\rangle\big) \tag{4.8}$$

then eqn (4.6) for the collective motion becomes non-linear because the Hartree–Fock Hamiltonian itself depends on the density.

Consider the motion of a typical nucleon inside the nucleus. Its angular momentum will be

$$\mathcal{I}_{\mathrm{N}}\, \omega_{\mathrm{N}} \simeq m_{\mathrm{N}} \langle r^2 \rangle\, \omega_{\mathrm{N}} \simeq \hbar \tag{4.9}$$

If we consider a typical collective rotation involving about 10 per cent of the nucleons then the collective angular momentum will be

$$\mathcal{I}_{\mathrm{c}}\, \omega_{\mathrm{c}} \simeq \tfrac{1}{10} A m_{\mathrm{N}} \langle r^2 \rangle\, \omega_{\mathrm{c}} \simeq \hbar \tag{4.10}$$

Hence the ratio of the individual nucleon orbital period to the collective nuclear period will be

$$T_{\mathrm{N}}/T_{\mathrm{c}} \simeq 10/A \simeq 1/20, \tag{4.11}$$

i.e. in the heavy nuclei the individual nucleons will typically execute about 20 revolutions for each rotation of the nuclear density. We are thus led to the adiabatic approximation (sometimes referred to as the Born–Oppenheimer approximation) in which we consider the changes in the nuclear densities to occur so slowly that at any instant of time a nucleon sees a static Hartree–Fock field. We then treat the time-dependence of the nuclear density in first-order perturbation theory

$$\rho(t) = \rho^{(0)} + \rho^{(1)}(t). \tag{4.12}$$

Inserting the form (4.12) in eqn (4.6) we have

$$i\hbar\frac{\mathrm{d}\rho_{ia}^{(1)}(t)}{\mathrm{d}t} = (\epsilon_a - \epsilon_i)\,\rho_{ia}^{(1)}(t) + \sum_{bj}(V_{ibaj}\,\rho_{jb}^{(1)} - V_{ijab}\,\rho_{bj}^{(1)}), \tag{4.13}$$

where, since we are dealing with a one-body density matrix, the transition density $\rho_{\alpha\beta}^{(1)}$ corresponds to the excitation of a single particle from the state α to the state β. In the static limit this means that the original state must be occupied in the Hartree–Fock Slater determinant (such states we shall label $i, j, k \ldots$), while the final state must be unoccupied (such states we shall label $a, b, c \ldots$). The occurrence of non-vanishing transition densities ρ_{bj} indicates that the ground state is no longer a single Slater determinant but does indeed contain many-body correlations. Such a transition density is frequently referred to as a 'backward-going' contribution, since it is equivalent to the transition of a nucleon from a state outside the static Fermi sea, which owing to the correlations, is now occupied into a hole in the static Fermi sea (see Fig. 4.1).

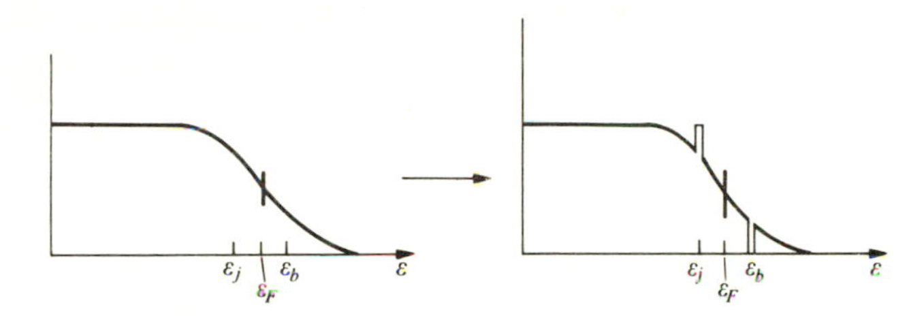

FIG. 4.1. A 'backward-going' configuration, i.e. a hole–particle state excited from a correlated configuration.

Assuming that the transition densities are of the form

$$\begin{aligned}\rho_{ia} &= \mathrm{e}^{iEt/\hbar}\,X_{ia},\\ \rho_{ai} &= \mathrm{e}^{-iEt/\hbar}\,Y_{ia},\end{aligned} \tag{4.14}$$

then eqn (4.13) and its complex conjugate may be written

$$\begin{aligned}\{E - (\epsilon_a - \epsilon_i)\}\,X_{ia} &= \sum_{bj}\{V_{ibaj}\,X_{jb} + V_{ijab}\,Y_{jb}\},\\ \{-E - (\epsilon_a - \epsilon_i)\}\,Y_{ia} &= \sum_{bj}\{V_{ibaj}\,Y_{jb} + V_{ijab}\,X_{jb}\}.\end{aligned} \tag{4.15}$$

These now form a set of linear algebraic equations which, given the residual interaction V and single-particle energies ϵ, can be solved numerically for the eigenenergies of collective motion E and the density-matrix coefficients X and Y.

So that we may see the essential qualitative features of the solution of eqns (4.15) let us consider a model interaction which is separable, i.e.

$$\langle \alpha\beta \mid V \mid \gamma\delta \rangle = \lambda D_{\alpha\gamma} D_{\beta\delta}. \tag{4.16}$$

First, let us assume there are no backward-going contributions (the Tamm–Dancoff approximation), then $Y = 0$ and

$$X_{ia} = ND_{ia}/\{E - (\epsilon_a - \epsilon_i)\}, \tag{4.17}$$

where N is the constant

$$N = \lambda \sum_{ia} D_{ia} X_{ia}. \tag{4.18}$$

Combining eqns (4.18) and (4.17) yields the dispersion relation

$$1/\lambda = \sum_{ai} D_{ia}^2/\{E - (\epsilon_a - \epsilon_i)\}, \tag{4.19}$$

which has the graphical solution illustrated in Fig. 4.2. Thus we see that for λ negative, i.e. an attractive residual interaction between the valence and core

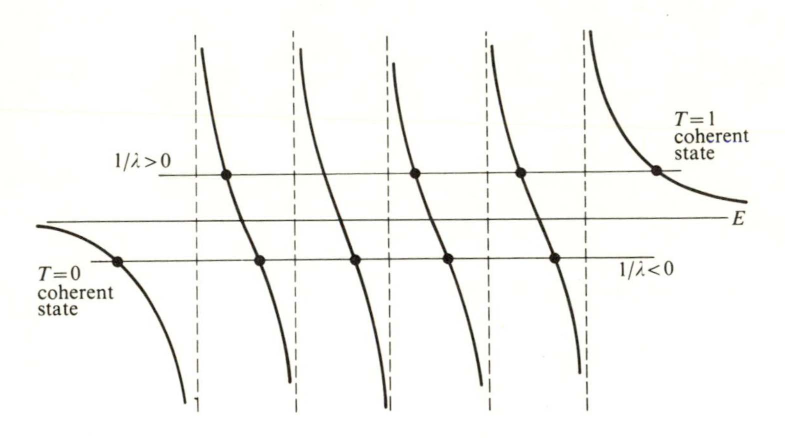

FIG. 4.2. A graphical solution of eqn (4.19).

particles, there is a collectively depressed state. If λ is positive, i.e. a repulsive residual interaction, then a state is collectively raised in excitation. As the strength of the interaction (i.e. λ) increases these collective enhancements increase without limit.

Following exactly the same procedure but with the backward-going contributions present (the particle–hole random-phase approximation (RPA)) and recognizing that $\lambda \sum_{ia} D_{ia}(X_{ia} + Y_{ia})$ is a constant, we obtain the dispersion relation

$$\frac{1}{\lambda} = \sum_{ai} \frac{2\, D_{ai}^2\, (\epsilon_a - \epsilon_i)}{(\epsilon_a - \epsilon_i)^2 - E^2}, \tag{4.20}$$

which has the graphical solution illustrated in Fig. 4.3. We see that the collectively raised state again rises without limit as λ increases. However, the collectively

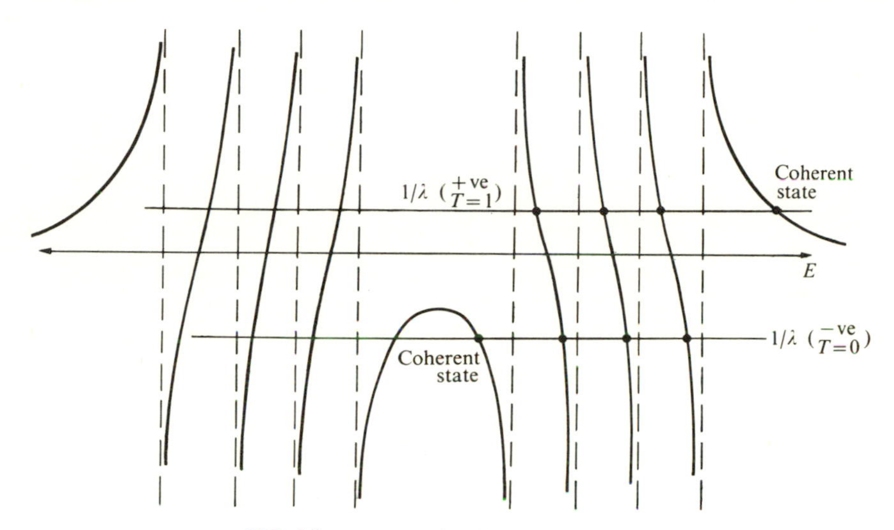

FIG. 4.3. A graphical solution of eqn (4.20).

lowered state develops a complex eigenvalue if the strength λ becomes too negative.

We may interpret these results as follows:

1. For $T = 1$ matrix elements, i.e. proton (neutron) valence particles interacting with proton (neutron) hole states, the residual interaction matrix elements are usually positive and the collective state is raised in energy. Physically, we see this as a vibration of the proton density out of phase with the neutron density. Since the neutron matter and proton matter attract each other this costs energy, and we would interpret such a collective excitation as typifying the giant dipole resonance seen, for example, in ^{208}Pb (Fig. 4.4).
2. For $T = 0$ matrix elements, i.e. neutron (proton) valence particles interacting with proton (neutron) hole states, the residual interaction is usually attractive and the collective state is lowered in energy. The low-lying octupole state in ^{208}Pb can be interpreted as such a collective excitation.

As the strength of the interaction increases (relative to the single-particle energy spacings) this collective state falls in energy until it passes through the ground state and eqn (4.20) yields complex eigenvalues. This implies that there has been a phase transition in the ground state which has gone from a spherically symmetric state supporting vibrations to a permanently deformed state which may or may not support collective vibrations but which certainly will support lower-lying collective rotations. In this case the static spherical Hartree–Fock Hamiltonian is no longer a good starting point for a perturbation calculation.

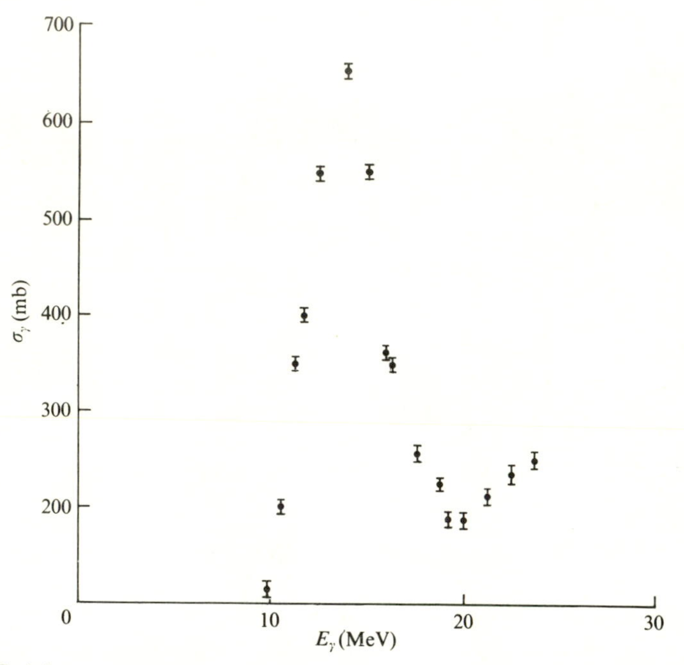

FIG. 4.4. Photon absorption cross-section versus photon energy for ^{208}Pb.

In the case of a deformed mean field we assume that the static Hartree–Fock Hamiltonian gives a good zero-order description of the nucleus in a body-fixed frame of reference in which the density is a function of coordinates x', y', and z'. For a nucleus rotating about the x-axis these coordinates are related to those measured in a laboratory fixed frame by

$$\begin{aligned} x' &= x, \\ y' &= y \cos \omega t + Z \sin \omega t, \\ z' &= -y \sin \omega t + Z \cos \omega t. \end{aligned} \tag{4.21}$$

Assuming that the rotation is slow, so that the relative spacing between particles is unaffected by this change of coordinate frame, the potential-energy contribution to the Hamiltonian is unchanged. However, the kinetic energy is affected,

$$\mathcal{H}(\mathbf{p}, \mathbf{r}) = \mathcal{H}(\mathbf{p}', \mathbf{r}') - \omega \sum_i (y_i' p_{z_i}' - z_i' p' y_i), \tag{4.22}$$

where the momenta are given by

$$\mathbf{p}_i' = \partial T / \dot{\mathbf{r}}_i, \tag{4.23}$$

Using eqns (4.21), eqn (4.22) reduces to

$$\mathcal{H}(\mathbf{p}, \mathbf{r}) = \mathcal{H}(\mathbf{p}', \mathbf{r}') - \omega L_{x'}, \tag{4.24}$$

where L_x is the angular momentum operator about the x-axis. For particles with spin $L_x \to J_x$ and hence transforming to a frame which is at rest with respect to the mean density yields the effective Hamiltonian

$$\mathcal{H}_\omega = \mathcal{H} - \omega J_x. \tag{4.25}$$

We can now use time-independent Hartree–Fock theory with this modified Hamiltonian. Eqns (3.49) for the single-particle states now become

$$(T + U)\, \psi_i - \omega j_x\, \psi_i = \epsilon_i\, \psi_i. \tag{4.26}$$

Using the solutions $\psi_i^{(0)}$ for $\omega = 0$ as a complete set of states we may express the solutions of eqn (4.26) in the form

$$\psi_i = \psi_i^{(0)} + \sum_m C_{mi}\, \psi_m^{(0)}. \tag{4.27}$$

whence

$$\begin{aligned} (\epsilon_m - \epsilon_i)\, C_{mi} + \sum_{nj} (\langle jm \,|\, V \,|\, ni \rangle C_{nj} + \langle mn \,|\, V \,|\, ij \rangle C_{nj}^\star) &= \omega \langle m \,|\, k_x \,|\, i \rangle, \\ (\epsilon_m - \epsilon_i)\, C_{mi}^\star + \sum_{nj} (\langle ij \,|\, V \,|\, mn \rangle C_{nj} + \langle ni \,|\, V \,|\, jm \rangle C_{nj}^\star) &= \omega \langle i \,|\, j_x \,|\, m \rangle. \end{aligned} \tag{4.28}$$

Our collective states are now Slater determinants of the single-particle states (4.27), where the coefficients C_{ij} are given by eqns (4.28).

Forming $E = \langle \psi \,|\, \mathcal{H} \,|\, \psi \rangle / \langle \psi \,|\psi \rangle$ the linear terms in C_{ij}s vanish since the $\psi_i^{(0)}$ are the solutions of the static Hartree–Fock equations. Keeping only the quadratic terms in the C_{ij} we have

$$\begin{aligned} E = E_0 + \sum_{mi} (\epsilon_m - \epsilon_i)\, |\, C_{mi}\, |^2 + \tfrac{1}{2} \sum_{imnj} (\langle ni \,|\, V \,|\, jm \rangle + \langle ij \,|\, V \,|\, mn \rangle) \\ (C_{nj}^\star\, C_{mi} + C_{nj}\, C_{mi}^\star). \end{aligned} \tag{4.29}$$

Interpreting the change in energy as due to rotational kinetic energy and defining a corresponding moment of inertia $\mathscr{I}_x$ by

$$E = E_0 + \tfrac{1}{2}\,\mathscr{I}_x\,\omega^2 \tag{4.30}$$

we have

$$\mathscr{I}_x = \frac{1}{\omega}\sum_{mi}(\langle m\,|j_x|\,i\rangle C^{\star}_{mi} + \langle i\,|j_x|\,m\rangle C_{mi}). \tag{4.31}$$

In the limit of extremely weak residual coupling, eqns (4.28) have the approximate solution

$$C_{mi} = \omega\langle m\,|j_x|\,i\rangle/(\epsilon_m - \epsilon_i), \tag{4.32}$$

whence we obtain the cranking-model expression for the moment of inertia

$$\mathscr{I}_x = 2\sum_{mi}|\langle m\,|j_x|\,i\rangle|^2/(\epsilon_m - \epsilon_i). \tag{4.33}$$

We could have written this result immediately from perturbation theory. Our Hamiltonian is (eqn (4.25))

$$\mathcal{H}_\omega = \mathcal{H} - J_x\,\omega, \tag{4.34}$$

whence

$$E = E_0 + \langle\Psi_0\,|J_x\omega|\,\Psi_0\rangle + \sum_i \omega^2\,|\langle\Psi_0\,|J_x|\,\Psi_i\rangle|^2/(E_0 - E_i + \ldots$$

$$\simeq E_0 + \tfrac{1}{2}\,\mathscr{I}_x\omega^2 \ldots$$

and

$$\mathscr{I}_x = 2\sum_i \frac{|\langle\Psi_0\,|J_x|\,\Psi_i\rangle|^2}{E_0 - E_i} = 2\sum_{mi}\frac{|\langle m\,|j_x|\,i\rangle|^2}{\epsilon_m - \epsilon_i}. \tag{4.36}$$

This is often referred to as the rigid-body value of the moment of inertia, because the neglect of the residual interaction in eqns (4.28) is equivalent simply to cranking round the static Hartree–Fock wavefunction while neglecting the changes in the self-consistent mean field induced by the rotations.

We notice that there is no contribution from the summation in (4.36) from states which are members of closed spherical shells since j_x is a linear combination of the angular momentum z-component raising and lowering operators. In a closed-shell configuration the states that can be reached by the action of j_x are already occupied, and hence the contributions must be zero. Thus we see that the microscopic model also predicts that only the valence particles contribute to the collective rotation.

The moments of inertia calculated in the cranking-model approximation are typically 5 times larger than those required to explain the rotational spectra observed in the heavy nuclei. The single-particle states and energies used in computing (4.36) may either be those for a spherically symmetric shell model or for a deformed Nilsson model. It is assumed that the number of terms required to give the dominant contribution to the summation will be minimized if the self-consistent Hartree–Fock states are used.

Yet another view of the problem set above is to view it as a variational calculation to find the best determinantal wavefunction for the Hamiltonian $\mathcal{H}$ (the Hartree–Fock problem), which, since the mean field is not spherically symmetric, will not be an eigenstate of angular momentum. However, in order to describe a rigid rotation about the x-axis it is necessary to have that component of angular momentum conserved. Hence we are faced with a constrained variational calculation (the Euler–Lagrange problem), where in (4.34) ω plays the role of a Lagrange multiplier.

4.3. Idealized residual interactions

The extreme individual particle model at first sight seems to play down the role of the residual interaction between nucleons as an originator of correlated motion, i.e. one deals with the zero-order Hartree–Fock Hamiltonian or assumes that the residual interaction is very weak. We have seen that this is not strictly true, since strong two-body correlations were essential in order to achieve a non-singular effective interaction (eqn (3.63)) from which the mean field could be generated. Thus the objects with which we are now dealing are not strictly individual nucleons but are nucleons coupled to each other via strong two-body correlations and an all-important mean field, they are quasi-particles. We shall now investigate how the residual interaction can lead to a further dressing of the quasi-particles, resulting in large-scale correlations in the nuclear motion.

Instead of seeking approximate solutions to a supposedly exact Hamiltonian we shall now consider various approximate Hamiltonians for which the Schrödinger equation can be solved exactly, in order to provide us with a guide to the nature of the correlations to be expected.

Given a residual interaction $V(r_{12})$ between the quasi-particles we can make a multipole expansion†

$$V(r_{12}) = \sum_{l} F_l(r_1, r_2) P_l(\cos\theta_{12}). \tag{4.37}$$

$P_l(\cos\theta_{12})$ versus θ_{12} is plotted in Fig. 4.5, and we see that the width of the dominant first peak falls rapidly with increasing l, i.e. low l values correspond to

† The multipole components of a force must not be confused with the partial wave components. The angle θ_{12} in eqn (4.37) is between the vectors $\mathbf{r}_1$ and $\mathbf{r}_2$, while in a partial wave decomposition the angles Ω_r are those which describe the orientation of the relative vector $\mathbf{r}_{12}$.

the longest-range components of the force. Typically a given multipole is characterized by a correlation angle $\sim 1/l$, i.e. a maximum range $r_l \lesssim R/l$, where R is the radius of the nucleus. In establishing the mean field which is the result of the interaction of a given nucleon with all the other nucleons in the nucleus we might expect the dominant contributor to be the long-range part of the interaction, i.e. the low multipoles. We shall assume that the low-multipole interactions have been exhausted in the establishment of the mean field, and we shall consider residual interaction with multipoles corresponding to $l > 2$.

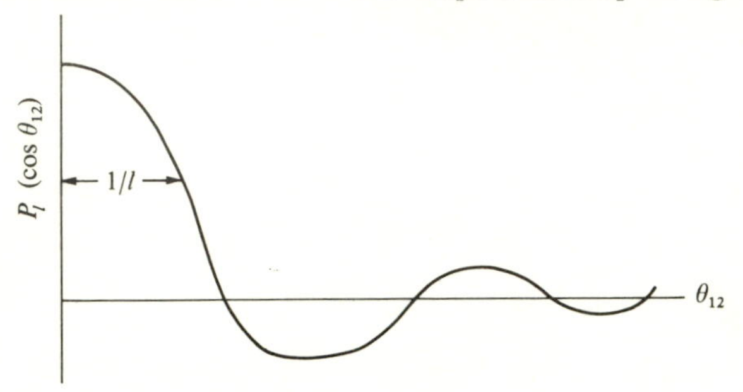

FIG. 4.5. The Legendre polynominal P_l (cos θ_{12}) as a function of θ_{12}.

4.3.1. The quadrupole–quadrupole interaction

We consider first the quadrupole moment, which may be written

$$V_Q = \sum_{ij} F_2(r_i, r_j) P_2(\cos\theta_{ij})$$

$$= \sum_{ij} \frac{4\pi}{5} F_2(r_i, r_j) \sum_m Y_{2m}(\theta_i, \phi_j)\, Y^\star_{2m}(\theta_j, \phi_j). \tag{4.38}$$

Thus our model Hamiltonian is of the form

$$\mathcal{H} = \mathcal{H}_{HF} + V_Q. \tag{4.39}$$

The mean field, being due to the $l = 0$ multipole, will be spherically symmetric, which for ease of analysis we shall assume to have the form of a harmonic oscillator potential. Further, we shall assume that the radial dependence of the residual interaction factorizes, thus

$$\frac{4\pi}{5} F_2(r_i, r_j) = -\chi r_i^2\, r_j^2. \tag{4.40}$$

Such a residual interaction is known as a quadrupole–quadrupole force. The contribution of such a force to the Hartree–Fock mean field will be of the form

$$\Delta U = -\chi \sum_m \int \rho(\mathbf{r}_2)\, r_2^2\, Y^\star_{2m}(\theta_2, \phi_2) \mathrm{d}\mathbf{r}_2\; r_1{}^2\; Y_{2m}(\theta_1, \phi_1). \tag{4.41}$$

Note that if the density is spherically symmetric the integral in (4.41) vanishes. However, if the density has a non-vanishing quadrupole moment, then the contribution to the mean field U has a collective quadrupole distortion and the Nilsson model becomes a self-consistent solution to (4.39).

We can now rewrite eqn (4.39) as

$$\mathcal{H} = \sum_{i=1}^{A} (p_i^2 + r_i^2) - \sum_{i>j} \sum_{m} \chi r_i^2 \, r_j^2 \, Y_{2m}^{\star} (\Omega_i) \, Y_{2m} (\Omega_j), \tag{4.42}$$

where we have set $m = \omega = \hbar = 1$. Defining the usual oscillator quanta creation and annihilation operators

$$\begin{aligned} a_i^{\dagger} &= \mathbf{r}_i - \mathrm{i}\mathbf{p}_i, \\ a_i &= \mathbf{r}_i + \mathrm{i}\mathbf{p}_i, \end{aligned} \tag{4.43}$$

and the quadrupole operators

$$q_{mi} = r_i^2 \, Y_{2m} (\Omega_i), \tag{4.44}$$

eqn (4.42) becomes

$$\mathcal{H} = \sum_i a_i^{\dagger} \, a_i - \chi \sum_{m,i,j} (-1)^m \, q_{mi} q_{-mj}.\dagger \tag{4.45}$$

Defining total angular momentum and quadrupole operators, namely,

$$\mathbf{L} = \sum_i \mathbf{l}_i \tag{4.46}$$

and

$$Q_m = \sum_i q_{mi}, \tag{4.47}$$

we can write

$$\mathcal{H} = \sum_i a_i^{\dagger} \, a_i - \chi \sum_m (-1)^m \, Q_m \, Q_{-m}. \tag{4.48}$$

Since the oscillator field is spherically symmetric it is possible to find eigenstates of the oscillator mean field which are simultaneously eigenstates of L^2 (the spherical shell model angular-momentum coupling problem). It is possible also to define an operator C,

$$C = \frac{1}{36} \left\{ 3L^2 + \sum_m (-1)^m \, Q_m \, Q_{-m} \right\}, \tag{4.49}$$

† The phase factor $(-1)^m$ arises from $Y_{jm} = (-1)^m \, Y_{j\,-m}$.

which commutes with the eight operators $L_{\pm} = L_x \pm iL_y, L_z, Q_m (m = -2, -1, 0, 1, 2)$ and hence with $L^2 = L_+ L_- + L_z{}^2$. In terms of this operator eqn (4.48) becomes

$$\mathcal{H} = \sum_i a_i{}^\dagger a_i - 36\chi C + 3\chi L^2. \qquad (4.50)$$

The eigenstates of $\Sigma a_i{}^\dagger a_i$ can be chosen to have good L^2, the term $-36C$ does not destroy this condition since C commutes with L^2. However, the $-36C$ term does produce a multipole distortion of the mean field through the quadrupole operators, i.e. the eigenstates of $\Sigma a_i{}^\dagger a_i - 36C$ are linear combinations of Nilsson-like states having good angular momentum. The operator $3L^2$ then simply shifts the energies of the levels by

$$\Delta E = 3\chi L(L+1)\hbar^2, \qquad (4.51)$$

i.e. a collective rotational band is generated with an effective moment of inertia

$$\mathcal{I}_{\text{eff}} = \tfrac{1}{6}\chi. \qquad (4.52)$$

In more mathematical language $L_\pm$, L_z, and the Q_m form a Lie group (the group SU_3) for which C is the Casimir operator. The eigenstates of (4.50) can then be labelled according to irreducible representatives of this group. The spectra of such states exhibit rotational bands (Irvine 1972).

4.3.2. Pairing interactions (Lane 1964)

We have seen that a quadrupole–quadrupole residual interaction is compatible with collective nuclear rotations. To what extent the effect of this type of residual interaction is already exhausted in producing a deformed mean field requires detailed numerical calculations. However, we may look on these interactions as probably being quite small in all but the most deformed (i.e. the heaviest) nuclei.

Turning now to higher multipoles of the residual interaction we see that these are of extremely short range. Hence we would expect the matrix elements of such interactions to be significant only between states which have the maximum spatial overlap one with the other. In dealing with the interaction between neutrons and protons this would clearly be the case where the neutron and proton are in exactly the same spatial state. However, in the heavy nuclei there is a considerable neutron excess (see section 3.2.1.), and since the residual interaction is effective only near the Fermi surface, where neutrons and protons are filling different spatial states, these interactions are heavily suppressed. Turning to identical particles the Pauli exclusion principle forbids these being in the same state. The overlap is now a maximum between time-reversed pairs of states, i.e. in j–j coupling between ψ_{jm} and $\psi_{j\text{-}m}$, in a momentum representation between $\psi_{k\sigma}$ and $\psi_{-k-\sigma}$, or in the Nilsson model between $\psi_{k\pi}$ and $\psi_{-k\pi}$, etc. Pairing theory is based on the assumption that the correlations between such pairs induced by the residual interaction are dominant.

To simplify the analysis we shall consider first the special case where all the particles are in degenerate single-particle states and where, in an obvious notation, the pairing matrix elements are all constant

$$\langle i, -i \mid V \mid j, -j \rangle = -G. \tag{4.53}$$

Assume that there are Ω pairs of states populated by N particles. We shall define the seniority s of a state as the number of unpaired particles. Clearly, eigenstates of the pairing interaction can be labelled by their seniority since this interaction scatters pairs of particles from one paired state into another paired state and hence does not destroy the seniority.

Obviously for N even the seniority is always even, while for N odd the seniority is always odd. Let us first consider even-N systems in seniority-zero states. The number of such configurations is $\Omega\,!/(\Omega - \frac{N}{2})\,!\,(\frac{N}{2})\,!$ For $N = 2$ there are thus Ω seniority-zero states and the Hamiltonian matrix is simply of order $\Omega \times \Omega$ and of the form

$$\mathcal{H} = -G \begin{bmatrix} 1 & 1 & 1 & . & . & . & . & . \\ 1 & 1 & . & . & . & . & . & . \\ 1 & . & . & . & . & . & . & . \\ . & . & . & . & . & . & . & . \\ 1 & . & . & . & . & . & . & 1 \end{bmatrix} \tag{4.53}$$

The eigenenergies of (4.53) are easily evaluated. There are found to be $(\Omega - 1)$ eigenstates with eigenvalue $E = 0$ and a single collectively depressed state with eigenvalue $E_0(2)$,

$$E_0(2) = -G\Omega; \tag{4.54}$$

the corresponding eigenfunction is

$$\psi_0(2) = \frac{1}{\sqrt{\Omega}} \begin{bmatrix} 1 \\ 1 \\ 1 \\ . \\ . \\ . \\ 1 \end{bmatrix}. \tag{4.55}$$

Note that this is truly a collective effect depending not simply on the strength G of the residual interaction but on the fact that many states act co-operatively.

All the seniority-2 states are degenerate with the $(\Omega - 1)$ seniority-zero states with $E = 0$, since for $N = 2$ there are no paired configurations of seniority-2 in which the residual pairing interaction can have an effect.

Similarly in a four-particle system the seniority-4 states all have energy $E = 0$. The seniority-2 states will have a spectrum similar to that of the seniority-zero two-particle states, except that the number of available paired states will be reduced by 2, i.e. the two unpaired particles will block two pairs of states, hence there will be a single collectively depressed seniority-2 state at $E_2(4) = -G(\Omega-2)$, the remainder having energy $E = 0$. There will be such a collectively depressed state for each of the $(\Omega-1)$ ways of blocking two states at a time. We can now obtain the energy of the seniority-zero ground state by computing the trace of the Hamiltonian and subtracting the total energy shift already calculated. The trace of the Hamiltonian is simply the number of four-particle states $\frac{1}{2}\,\Omega(\Omega-1)$ times the diagonal matrix element $-2G$ (the factor 2 arises since the diagonal element occurs twice in the sum $\sum_{ij} \langle i-i \mid V \mid j-j \rangle$). Thus

$$\mathrm{Tr}\mathcal{H} = -\,\Omega(\Omega-1)G \tag{4.56}$$

and

$$E_0(4) = -\,\Omega(\Omega-1)G + (\Omega-1)(\Omega-2)G = -\,2\,(\Omega-1)G. \tag{4.57}$$

Continuing in this manner we arrive at the general expression for the collectively depressed seniority-s states of an N-body system

$$E_s(N) = -\tfrac{1}{4}\,G(N-s)(2\Omega-N-s+2). \tag{4.58}$$

We can quickly check that this formula is also valid for odd-mass systems. Clearly for $N = 1$ we have $s = 1$ only and the 2Ω states are all degenerate with $E_1(1) = 0$. For $N = 3$ we have $s = 1$ or 3. The states with $s = 3$ are clearly all completely unpaired and are degenerate with $E_3(3) = 0$, while the $s = 1$ states can be thought of as similar to the $N = 2$ system with the third particle simply blocking one state, thus $E_1(3) = -\,G(\Omega-1)$, etc.

The typical excitation spectrum of neighbouring even- and odd-mass systems is sketched in Fig. 4.6. We note that there is always a gap Δ of order ΩG between

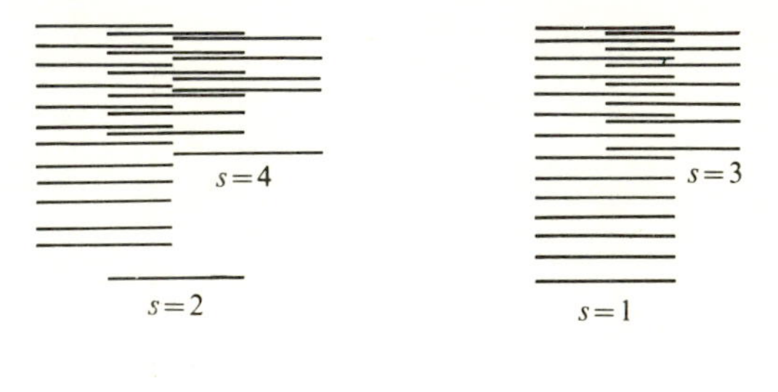

FIG. 4.6. Typical seniority spectra for (a) even-mass nuclei; (b) odd-mass nuclei.

the seniority-zero ground state and the first seniority-2 state of the even-mass system, while no such gap appears in the odd-mass spectrum. Secondly, there is a difference in the ground-state binding energies of at least ΩG. We can thus tentatively make the connection between the gap parameter and the phenomenological pairing term in the semi-empirical mass formula (section 3.2.1). Since the neutrons and protons pair separately, we see that by comparing neighbouring isotopes and isotones we can estimate the neutron- and proton-pairing constants G_N and G_Z respectively.

While the odd–even mass spectral difference is evident in lighter nuclei it is not apparent in the heavy nuclei $A \gtrsim 220$. For example, in Fig. 4.7 we compare

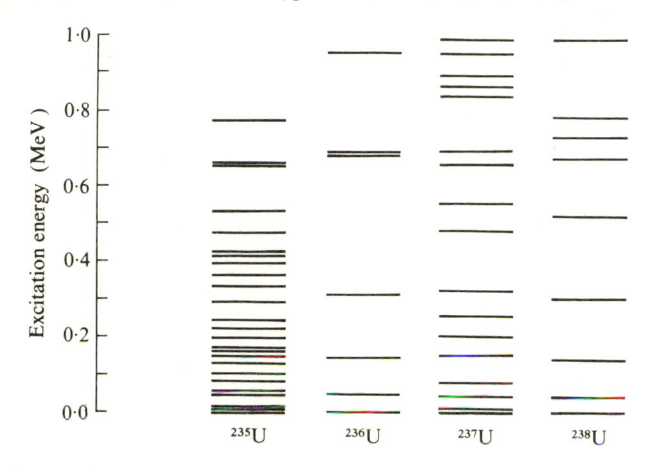

FIG. 4.7. Density of low-lying states in even- and odd-mass uranium isotopes.

the spectra of neighbouring uranium isotopes, and while the ground states exhibit the usual odd–even mass effect there is no apparent large gap in the excitation spectra of the even isotopes. However, if we look more closely at the states below 1 MeV in the even-mass spectrum we see that they are either collective rotational states or, in the case of the 1^- state, a collective vibrational state. It is the fact that the intrinsic ground state of the even-mass isotopes has been collectively depressed that allows us to see clearly the collective rotations and vibrations built on this highly correlated configuration. Had no pairing gap existed the low-lying spectra of even-mass nuclei would have been so cluttered up with intrinsic excitations that the identification of the simple collective rotational bands might never have been made.

To stress the collective nature of these depressed states let us consider ground-state electromagnetic transitions and in particular E2 transitions. First, since all seniority-zero states have $J = 0$ we cannot have an E2 transition within the seniority-zero spectrum. Secondly, electromagnetic transitions involve one-body operators and hence ground-state transitions can only come from seniority-2 states. Let us consider the E2 transition between the lowest seniority-2 state and the ground state. Because of the selection rules discussed above, we have

$$|\langle s = 2 \,|\, Q \,|\, s = 0 \rangle|^2 \simeq \langle s = 0 \,|\, Q^2 \,|\, s = 0 \rangle. \tag{4.59}$$

Assuming that the valence particles are occupying a single j shell, i.e. $\Omega = \frac{1}{2}(2j+1)$, we find that

$$|\langle s=2\,|Q|\,s=0\rangle|^2 = \frac{N(\Omega - N/2)}{2\Omega(\Omega-1)} \sum_m \langle q_m^2 \rangle, \tag{4.60}$$

where q_m is the quadrupole operator of eqn (4.44). Now we see that $1/\Omega\ \Sigma\ \langle q_m^2 \rangle$ is nothing more than the square of the transition-matrix element for the two-particle system.† Thus for $N \ll \Omega$ we see that the transition rate is proportional to N and can thus be many times the single-particle value, i.e. there is a collective enhancement of the transition rate.

We now consider the more realistic situation in which the single-particle states are not all degenerate. For this we shall have to work in the occupation number representation, i.e. we define a set of creation and annihilation operators $a_i{}^\dagger$ and a_i which create or destroy a single particle in the state ϕ_i. Thus

$$\Phi_n = \frac{1}{\sqrt{A!}} \det \phi_i(\mathbf{x}_j) \rightarrow \prod_{i \leq n} a_i^\dagger\,|0\rangle, \tag{4.61}$$

where the vacuum state is defined by

$$a_i\,|0\rangle = 0, \text{ all } i. \tag{4.62}$$

To provide the antisymmetry of the many-fermion wavefunction these operators obey the fermion commutation relations

$$[a_i^\dagger, a_{i'}^\dagger]_+ = [a_i, a_{i'}]_+ = 0,\ [a_i^\dagger, a_{i'}]_+ = \delta_{ii'}. \tag{4.63}$$

The Hamiltonian with residual pairing interaction may now be written

$$\mathcal{H} = \sum_k \epsilon_k\, a_k^\dagger a_k - G \sum_{kk'} a_k^\dagger a_{\bar{k}}^\dagger a_{\bar{k}'} a_{k'}. \tag{4.64}$$

We shall seek a variational approach to the problem and consider the general seniority-zero trial wavefunction

$$\Psi = \prod_{k>0} (u_k + v_k\, a_k^\dagger a_{\bar{k}}^\dagger)\,|0\rangle \tag{4.65}$$

In order that Ψ be normalized we require that

$$u_k{}^2 + v_k{}^2 = 1, \tag{4.66}$$

so that only the v_ks are independent variational parameters. The trial function (4.65) is not an eigenstate of the number of particles, hence we shall carry out a

† The $1/\Omega$ factor simply arises from the normalization of the ground-state vector (see eqn (4.55)).

constrained variational calculation subject to the subsidiary condition that the mean number of particles is fixed at its true value, i.e.

$$\langle \Psi | \hat{N} | \Psi \rangle = \left\langle \Psi \left| \sum_k a_k^\dagger a_k \right| \Psi \right\rangle = A. \tag{4.67}$$

This introduces a Lagrange multiplier ϵ_F, and the variational equations are then

$$\frac{\partial}{\partial v_k} \langle \Psi | \mathcal{H} - \epsilon_F \hat{N} | \Psi \rangle = 0. \tag{4.68}$$

This is equivalent to a grand canonical averaging of observables, and statistical physics tells us that the results are subject to fluctuations of order $1/\sqrt{A}$, which is an acceptable level of accuracy in the heavy nuclei. ϵ_F has the usual interpretation as a chemical potential, i.e. the Fermi energy.

Inserting (4.64), (4.65), and (4.67) in eqns (4.68) we have

$$\frac{\partial}{\partial v_k} \left\{ \sum_k (\epsilon_k - \epsilon_F) v_k^2 - G \left(\sum_k u_k v_k \right)^2 \right\}$$

$$= 2(\epsilon_k - \epsilon_F)\, v_k - G\left(\sum_i u_i v_i \right) \left\{ u_k - v_k^2 / u_k \right\} = 0. \tag{4.69}$$

and substituting from eqn (4.66) we have the solutions

$$v_k^2 = \tfrac{1}{2} \left[1 - (\epsilon_k - \epsilon_F) \{ (\epsilon_k - \epsilon_F)^2 + \Delta^2 \}^{-\frac{1}{2}} \right] \tag{4.70}$$

and

$$u_k^2 = \tfrac{1}{2} \left[1 + (\epsilon_k - \epsilon_F) \{ (\epsilon_k - \epsilon_F)^2 + \Delta^2 \}^{-\frac{1}{2}} \right], \tag{4.71}$$

where

$$\Delta = G \sum_k u_k v_k \tag{4.72}$$

v_k^2 may be interpreted physically as the probability that the single-particle state ϕ_k is occupied in Ψ,

$$\langle \Psi | a_k^\dagger a_k | \Psi \rangle = v_k^2, \tag{4.73}$$

and the form of the occupation probability is sketched in Fig. 4.8. From eqn (4.66) we have the interpretation that u_k^2 is the probability that this state is unoccupied.

Substituting (4.70) and (4.71) in eqn (4.72) yields the dispersion relation

$$\Delta = \tfrac{1}{2} \Delta G \sum_k \{ (\epsilon_k - \epsilon_F)^2 + \Delta^2 \}^{-\frac{1}{2}}, \tag{4.74}$$

which has the two solutions, either $\Delta = 0$ or

$$\frac{2}{G} = \sum_k \{(\epsilon_k - \epsilon_F)^2 + \Delta^2\}^{-\frac{1}{2}}. \tag{4.75}$$

Note that the solution $\Delta = 0$ corresponds to $v^2 = 1$ for $\epsilon_k < \epsilon_F$ and $v_k{}^2 = 0$ for $\epsilon_k > \epsilon_F$, and this is simply the uncorrelated Hartree–Fock solution.

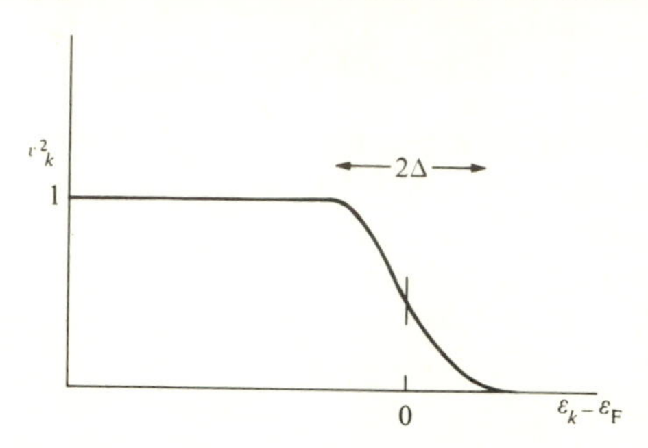

FIG. 4.8. The correlated paired ground state.

For further insight into the behaviour of our system we shall assume a constant smooth density of single-particle states ρ and convert the summation in (4.75) into an integral

$$\frac{2}{G} = \rho \int_{\omega}^{\omega} \frac{d\epsilon}{\sqrt{(\epsilon^2 + \Delta^2)}}, \tag{4.76}$$

where the residual interaction connects paired states within a range ω on either side of the Fermi surface. The solution of eqn (4.76) is

$$\Delta = \omega / \sinh(1/Gp). \tag{4.77}$$

For a weakly interacting system $G\rho \ll 1$ and

$$\Delta \simeq \omega \exp(-1/\rho G). \tag{4.78}$$

The importance of the above result lies in the sensitivity it shows to the density of single-particle states at the Fermi surface. We see from Fig. 3.18 (pp. 50 and 51) that this can be extremely sensitive to the deformation of the nucleus.

Let us now calculate the total contribution of the pairing energy to the nucleus. It is

$$E_\rho = \sum_k (\epsilon_k - \epsilon_F) v_k{}^2 - G\left(\sum_k u_k v_k\right)^2 - \sum_{\epsilon_k < \epsilon_F} \epsilon_k \simeq \rho(\omega^2 - \omega\sqrt{(\omega^2 + \Delta^2)}) \tag{4.79}$$

or, letting $\omega \to \infty$, we have

$$E_\rho \simeq -\rho\Delta^2. \tag{4.80}$$

In heavy nuclei we have seen that $\Delta \sim 0{\cdot}5-1$ MeV, while the density of single-particle levels may be deduced from the spectra of odd-mass nuclei and is typically $\sim 10-20$ states per MeV. Thus the total pairing energy is of order 10 MeV, which is small compared to a total binding energy of $\sim 1{\cdot}6-1{\cdot}8$ GeV.

Let us now consider seniority-1 states in the shell model. We consider an odd-mass nucleus of $(A + 1)$ nucleons where the low-lying states $\Psi_{jm}(A + 1)$ are thought of as being produced by adding a single particle to an even-mass fully-paired seniority-zero nuclear state $\Psi_0(A)$,

$$\Psi_{jm}(A + 1) = a_{jm}^{\dagger}\ \Psi_0(A). \tag{4.81}$$

However, the possibility of doing this must be weighted with the probability amplitude that the single-particle state ϕ_{jm} is not occupied in $\Psi_0(A)$, i.e. u_k. Similarly, we could have arrived at the same state by taking one particle away from the seniority-zero $(A + 2)$ particle ground state

$$\Psi_{jm}(A + 1) = a_{j\,-\,m}\ \Psi_0(A + 2), \tag{4.82}$$

where the possibility of doing this must be weighted with the probability that the state $\phi_{-\,k}$ is occupied in $\Psi_0(A + 2)$, i.e. v_k. The wavefunction (4.65) contains both the A and the $(A + 2)$ ground states, and provided $A \gg 1$ the occupation probabilities $v_k{}^2$ will not change greatly in going from the A to $(A + 2)$ ground state. Since we have no *a priori* method of choosing between (4.81) and (4.82) we shall assume that a normalized linear combination of these states is appropriate

$$\Psi_{jm} = (u_{jm}\ a_{jm}^{\dagger} - v_{jm}\ a_{j\,-\,m})\ \Psi_0, \tag{4.83}$$

and we denote the creation of this seniority-1 state by a single quasi-particle operator

$$\alpha_{jm}^{\dagger} = u_j\ a_{jm}^{\dagger} - v_j\ a_{j\,-\,m}. \tag{4.84}$$

It is trivial to prove that Ψ_0 is the vacuum of these quasi-particle operators

$$\alpha_{jm}\ \Psi_0 = 0, \text{ all } j, m; \tag{4.85}$$

furthermore, these quasi-particles are fermions

$$[\alpha_m, \alpha_m']_+ = [\alpha_m^{\dagger}, \alpha_m^{\dagger}{}']_+ = 0,\ [\alpha_m^{\dagger}, \alpha_m{}']_+ = \delta_{mm'}. \tag{4.86}$$

We can now rewrite our pairing Hamiltonian in terms of quasi-particle operators, and keeping only terms corresponding to non-interacting quasi-particles we find

$$\mathcal{H} = E_0 + \sum_k E_k\ \alpha_k{}^{\dagger}\ \alpha_k, \tag{4.87}$$

where the quasi-particle energies are given by

$$E_k = \surd\{(\epsilon_k - \epsilon_{\mathrm{F}})^2 + \Delta^2\}. \tag{4.88}$$

The ground state of the seniority-1 system is formed by adding a particle at the Fermi energy, whence

$$E_{k_0}(1) = E_0 + \Delta, \tag{4.89}$$

and the excitation of the state $\Psi_k(1)$ is given by

$$E_k(1) - E_{k_0}(1) \simeq (\epsilon_k - \epsilon_F)/\sqrt{\{(\epsilon_k - \epsilon_F)^2 + \Delta^2\}}. \tag{4.90}$$

i.e. the levels are closer together than they would be in a system without pairing. This is due to the fact that if a particle is added close to the Fermi surface it blocks this state from interacting with all the other paired states, and hence this state suffers a reduction in the pairing energy. However, if the odd particle is added to a level some distance from the Fermi surface there is little effect on the pairing energy. The form of the quasi-particle energies is illustrated in Fig. 4.9.

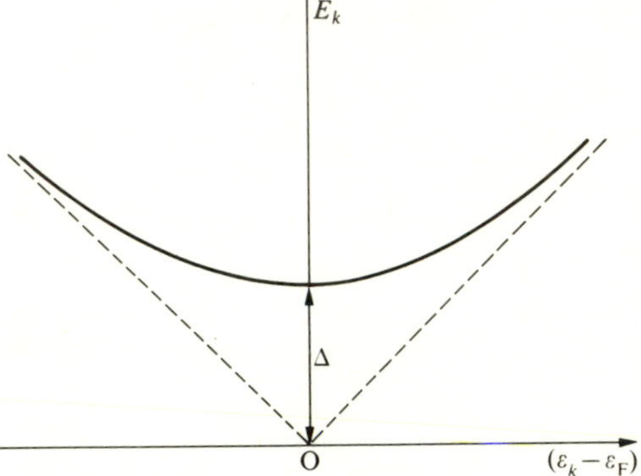

FIG. 4.9. The pairing quasi-particle energies. The dashed lines correspond to the uncorrelated energies.

Returning to the even-mass system the seniority-2 states are obtained by adding two quasi-particles to the seniority-zero ground state

$$\Psi_{k_1 k_2}(2) = \alpha_{k_1}^\dagger \alpha_{k_2}^\dagger \Psi_0 \tag{4.91}$$

and lies at an energy

$$E_{k_1 k_2}(2) = E_{k_1} + E_{k_2} = \sqrt{\{(\epsilon_{k_1} - \epsilon_F)^2 + \Delta^2\}} + \sqrt{\{(\epsilon_{k_2} - \epsilon_F)^2 + \Delta^2\}}, \tag{4.92}$$

and hence there is a gap of at least 2Δ between the ground state and first excited state of an even-mass system.

We must next consider how to improve the simple pairing theory outlined above. First, we cannot expect the interaction-matrix elements to be constant, i.e. we should generalize the Hamiltonian (4.64) as follows

$$\mathcal{H} = \sum_k \epsilon_k a_k^\dagger a_k - \sum_{kk'} G_{kk'} a_{k'}^\dagger a_{\bar{k}'}^\dagger a_{\bar{k}} a_k. \tag{4.93}$$

This has the consequence simply of making the gap parameter Δ state-dependent. The dispersion relation (4.74) now reads

$$\Delta_k = \tfrac{1}{2} \sum_{k'} G_{kk'} \Delta_{k'} \{(\epsilon_{k'} - \epsilon_F)^2 + \Delta_{k'}\}^{-\frac{1}{2}}, \tag{4.94}$$

which has to be solved numerically. Qualitatively, however, no new phenomena are introduced.

Secondly, we consider the effect of the residual interactions between the quasi-particles which have been neglected in our extreme pairing model (4.93). Such interactions must serve to break pairs of particles. Thus these interactions acting on a seniority-zero ground state produce seniority-2 states. Therefore we can define a creation operator which produces such excitations

$$Q_{jm}^{\dagger} = \sum_{k_1 k_2} \{ X_{k_1 k_2} (\alpha_{k_1}^{\dagger} \alpha_{k_2}^{\dagger})_m^j + y_{k_1 k_2} (\alpha_{k_1} \alpha_{k_2})_m^j \}, \tag{4.95}$$

where our notation implies that the pairs of quasi-particles are coupled to angular momentum j and component m. The second term in (4.95) would yield zero acting upon the unperturbed paired ground state which is the quasi-particle vacuum (see eqn (4.85)). However, if the non-pairing matrix elements can produce further correlations which destroy the *exact* seniority scheme then these quasi-particle destruction operators will play the same role as the 'backward-going' contributions of eqns (4.13) and (4.15). Clearly j must be even for identical particles; since the ground state is already antisymmetric, if we require the excited state

$$|jm\rangle = Q_{jm}^{\dagger} \,|\Psi_0\rangle \tag{4.96}$$

to be antisymmetric then the $Q_{jm}^{\dagger}$ must be symmetric (i.e. bose) operators. However, the $j = 0$ are simply the pairing terms, thus the first excitation should have $j = 2$. The quasi-particle quadrupole phonons will lie at an energy $E_2 \gtrsim 2\Delta$. The two-phonon state

$$|\Psi_4(2)JM\rangle = [Q_{jm}^{\dagger} Q_{jm}^{\dagger}]_M^J \,|\Psi_0\rangle \tag{4.97}$$

should lie at approximately 4Δ and because we have bose operators only the $J = 0^+, 2^+, 4^+$ triplet of states should exist. The exact degeneracy of the triplet will be split by the residual interaction between the quasi-particles. Thus we see that the non-pairing terms in the residual interaction lead to a vibrational spectrum. The importance of these terms will be relatively enhanced where the pairing correlations are weakest. We might expect this to occur where the density of states at the Fermi surface is low (see eqn (4.78)), which is near a closed shell.

We have now completed our survey of the dominant nuclear correlations, and we have shown how these collective features can arise through the effects of the residual interaction between valence nucleons. It is up to specific numerical calculations to show that these explanations can yield quantitative agreement with experiment. In Chapter 5 we shall pursue in more detail the complex question of collective nuclear rotations.

5

COLLECTIVE ROTATIONS

We have seen in Chapters 3 and 4 that both the extreme liquid-drop model and the individual-particle model can give rise to collective rotational spectra, and hence we may feel satisfied that a qualitative description can be given of the gross properties of the low-lying levels in the even–even heavy nuclei. In the present chapter we shall seek to examine the additional details which are required in order to bring the theory and the experiments into quantitative agreement (Belyaev 1959; Davidson 1968; Sorenson 1973).

In the periodic table there are two regions in which the nuclei exhibit large deformations and clear evidence of rotational bands: the rare-earth nuclei and the extremely heavy nuclei. In the rare-earth nuclei the rotational bands are studied up to very high spin states (typically $I \gtrsim 20\hbar$). Schematically, a typical situation would have the appearance of Fig. 5.1. The high spin states are populated by heavy-ion reactions, resulting in the production of the rare-earth nucleus in a high spin state together with the emission of some neutrons and the subsequent gamma radiation indicated in the Figure. We denote such a reaction

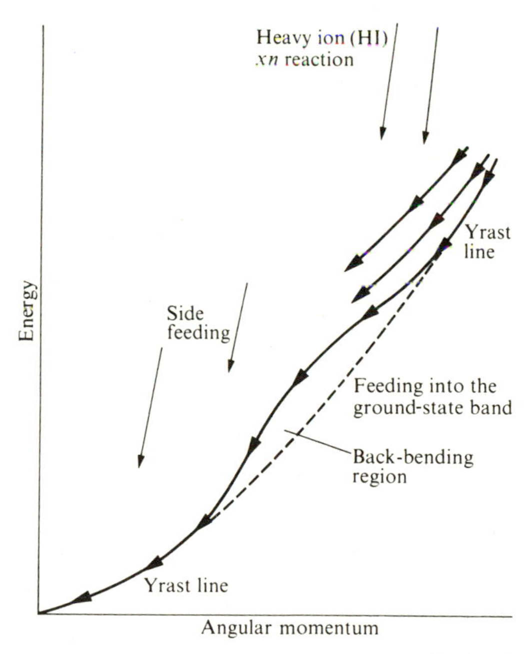

FIG. 5.1. Schematic population and decay of high spin states.

by (HI, xn, γ). Heavy ions are used because they can carry large quantities of angular momentum into the reaction even at quite modest energies.

In the case of heavy nuclei the situation is more complicated since the large Coulomb barriers between the heavy ions mean that (HI, xn, γ) reactions can only only occur at very high energies. However, at such high energies the fission lifetime is extremely short and experiments become exceedingly difficult. In the heavy nuclei we can turn the large Coulomb barriers to our advantage as follows. Heavy ions carry large Coulomb fields, there will be a range of energies for which the Coulomb barrier between the target and projectile will ensure that they do not come close enough for the nuclear forces to become operative but for which the target experiences a strong rapidly varying electromagnetic field due to the proximity of the projectile. Energy taken from this electromagnetic field can lead to the Coulomb excitation of the target. In this manner states up to the 20^+ level in ^{238}U have been identified.

There are two ways in which a detailed study of the observed rotational spectra show differences from the simple theories developed in Chapters 3 and 4. First, the moment of inertia is not constant as we go up the rotational band. Secondly, from the spacing of the low spin states the observed moment of inertia is considerably smaller than calculations with the cranking model would suggest. We shall tackle the latter problem first and consider the effects of pairing correlations on rotational spectra.

We begin by considering the Hamiltonian (4.25),

$$\mathcal{H}_\omega = \mathcal{H} - \omega J_x, \tag{5.1}$$

and write the operator J_x in terms of non-interacting quasi-particles,

$$J_x = \sum_{kk'} \langle k | j_x | k' \rangle a_k^\dagger a_{k'} = \sum_{kk'} \langle k | j_x | k' \rangle (u_k v_{k'} - u_{k'} v_k)\, \alpha_k^\dagger \alpha_{k'}^\dagger + \ldots, \tag{5.2}$$

where we have only kept the terms containing two creation operators because the ground state Ψ_0 is the quasi-particle vacuum, and hence the quasi-particle annihilation terms in (5.2) would not contribute to a *ground-state* rotational band. The operator J_x thus only connects the ground-state seniority-zero configuration to the two quasi-particle states. Thus we may write our trial wavefunction in the rotating frame as

$$\Psi_\omega = \left(1 + \sum_{ij} C_{ij} \alpha_i^\dagger \alpha_i^\dagger\right) \Psi_0 \tag{5.3}$$

and treat the C_{ij} as variational parameters in minimizing the expectation value of $\mathcal{H}_\omega$. In the approximation of non-interacting quasi-particles this yields

$$(E_i + E_j) C_{ij} = \omega (u_i v_j - v_i u_j) \langle i | j_x | j \rangle. \tag{5.4}$$

Substituting back into (5.3) for C_{ij} we thus obtain

$$\langle \Psi_\omega | \mathcal{H} - \omega J_x | \Psi_\omega \rangle / \langle \Psi_\omega | \Psi_\omega \rangle = E_0 + \tfrac{1}{2} \mathscr{I}\omega^2, \tag{5.5}$$

where the effective moment of inertia has the form

$$\mathscr{I} = \sum_{ij} |\langle i | j_x | j \rangle|^2 (u_i v_j - v_j u_i)^2 / (E_i + E_j). \tag{5.6}$$

This should be compared with the cranking-model expression (4.33). Clearly (5.6) yields a smaller calculated moment of inertia, since now the energy denominator contains the quasi-particle energies and is always $> 2\Delta$. Also all the factors u_i and v_i are less than unity. Indeed, numerical calculations suggest that the reduction in the moment of inertia due to pairing correlations suggested by (5.6) is too severe and the calculated moments of inertia are now some 20 per cent smaller than those observed. This discrepancy can be accounted for by including the seniority-breaking non-pairing terms in the residual interaction discussed at the end of Chapter 4. This requires that we keep the quasi-particle annihilation terms in eqn (5.2) and the quasi-particle interaction terms in eqn (5.4).

We now consider the variations in the moment of inertia as we go up a rotation band. The moment of inertia increases universally as we start up the band, and amongst the heavy nuclei this increase persists up to the highest members of the ground-state band so far observed. However, as mentioned before, it is difficult to excite very high angular momentum states in the heavy nuclei and very few rotation bands have been studied beyond the 14^+ member. In the rare-earth nuclei a large number of bands have been studied up to $\sim 20^+$ levels. In many cases it is observed that around the 14^+–16^+ members there is a sudden dramatic increase in the moment of inertia. In a few cases this is followed at the 18^+–20^+ members by a fall in the moment of inertia. These effects are represented schematically in Fig. 5.2. An increasing moment of inertia is observed as a compression of the rotational band.

It has become conventional to plot the moment of inertia deduced from the observed levels through

$$E_J = J(J+1)\hbar^2/2\mathscr{I} \tag{5.7}$$

against the square of the angular velocity ω. Classically, this angular velocity is defined as follows

$$J = \mathscr{I}\omega \tag{5.8}$$

and

$$\omega = \mathrm{d}E/\mathrm{d}J. \tag{5.9}$$

Quantum-mechanically the energy E_J is no longer a continuous function of J, and hence there must be some slight ambiguity in the definition of ω. If we

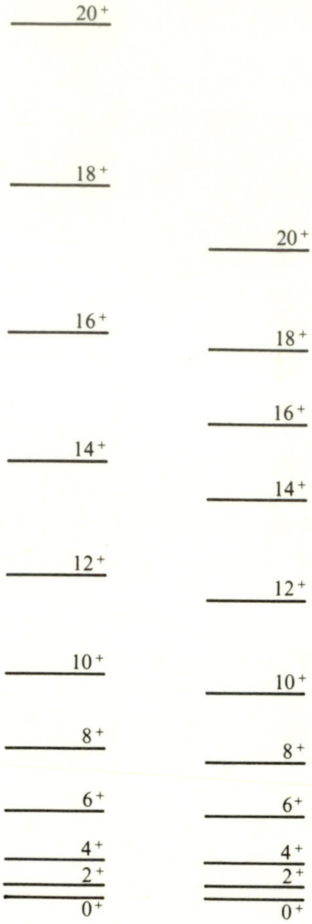

20⁺ 18⁺ 16⁺ 14⁺ 12⁺ 10⁺ 8⁺ 6⁺ 4⁺ 2⁺ 0⁺

20⁺ 18⁺ 16⁺ 14⁺ 12⁺ 10⁺ 8⁺ 6⁺ 4⁺ 2⁺ 0⁺

FIG. 5.2. A comparison of a rigid rotational band with a band showing the typical variations in moment of inertia observed in the rare-earth nuclei.

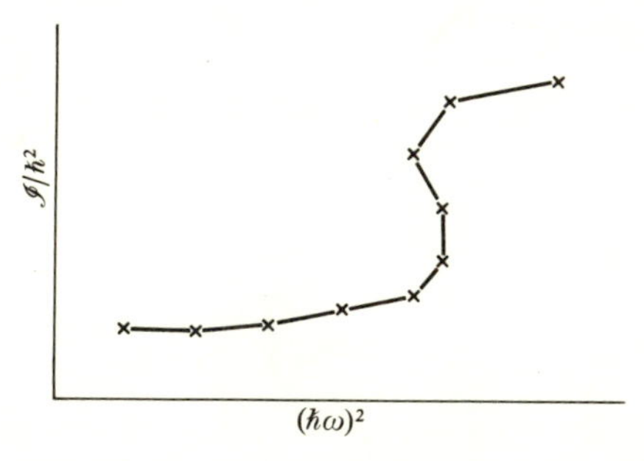

FIG. 5.3. Nuclear 'back-bending' corresponding to the band in Fig. 5.2.

follow the classical prescription then the best approximation we have to (5.9) is

$$\omega = (E_J - E_{J-2}) / 2\hbar^2. \tag{5.10}$$

We note that this is simply half the frequency of the emitted gamma decay in the $E_J \rightarrow E_{J-2}$ quadrupole decay, which is exactly what we would expect for a deformed nucleus with reflection symmetry such that a rotation through π radians brings the system back to its original shape. The resulting curves of $\mathscr{I}$ versus $(\hbar\omega)^2$ are plotted in Fig. 5.3, corresponding to the curves in Fig. 5.2, and we see the origin of the term 'back-bending' of nuclear moments of inertia. Fig. 5.4 shows the actual plots corresponding to observations on some heavy nuclei.

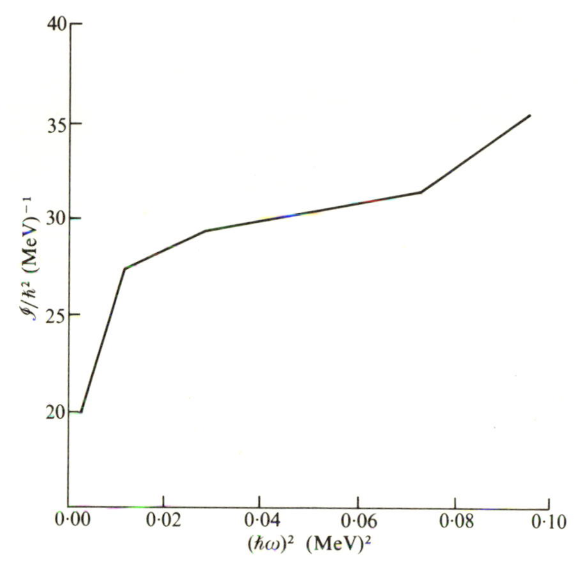

FIG. 5.4. The moment of inertia versus the square of the angular velocity for the ground state rotation bands of the heavy nuclei. From Fig. 2.5 we see that this is independent of mass number for the low-lying states.

Up to this point we have assumed complete adiabaticity, i.e. there is no coupling between the collective rotation and the mean field. Classically, of course, there are the well-known coriolis and centrifugal forces acting on a particle in a rotating frame of reference

$$F_{\text{cent}} = - m\boldsymbol{\omega} \times (\boldsymbol{\omega} \times \mathbf{r}),$$

$$F_{\text{cor}} = - 2m\boldsymbol{\omega} \times \dot{\mathbf{r}}. \tag{5.11}$$

For a constant angular velocity about the z-axis the centrifugal force can be described by a pseudopotential of the form

$$V_{\text{cent}} = - \tfrac{1}{2} m\, \omega^2 (x^2 + y^2). \tag{5.12}$$

The coriolis force is equivalent to a uniform magnetic field in the z-direction of magnitude $2m\omega$ and can be described by a vector potential

$$A_{\text{cor}} = m\boldsymbol{\omega} \times \mathbf{r}. \tag{5.13}$$

Since we have argued that the collective angular velocity ω is small compared with the mean nucleon velocity we can treat these additions to the mean field as perturbations in which the coriolis force appears in first order while the centrifugal force appears in second order in the wavefunction. †

We shall consider first the effect of the coriolis force in lowest-order perturbation theory. We shall assume that the unperturbed mean field is of a Nilsson type with single-particle states ϕ_k degenerate with their time-reversed partners ϕ_{-k}. The coriolis force will perturb these single-particle states, thus

$$\phi_k \rightarrow \psi_k = a_k \phi_k + \omega \sum_{k'} b_{kk'} \phi_{k'}, \tag{5.14}$$

where if the ψ_k are normalized to unity,

$$|a_k|^2 + \omega^2 \sum_{k'} |b_{kk'}|^2 = 1. \tag{5.15}$$

However, the coriolis force is odd under time-reversal, thus

$$\phi_{-k} \rightarrow \psi_{-k} = a_k \phi_{-k} - \omega \sum_{k'} b_{kk'} \phi_{-k'}. \tag{5.16}$$

Thus a typical pairing matrix element will become

$$\langle \phi_k \phi_{-k} | G | \phi_{k'} \phi_{-k'} \rangle \rightarrow \langle \psi_k \psi_{-k} | G | \psi_{k'} \psi_{-k'} \rangle$$

$$= G\Big(|a_k|^2 - \omega^2 \sum_{k''} |b_{k''k}|^2\Big)\Big(|a_{k'}|^2 - \omega^2 \sum_{k''} |b_{k''k'}|^2\Big). \tag{5.17}$$

However, the summations in (5.15) and (5.17) are over positive definite quantities and hence the factors

$$|a_k|^2 - \omega^2 \sum_{k'} |b_{k'k}|^2 \leqslant 1, \tag{5.18}$$

and in turn the pairing-matrix element in (5.17) is $\leqslant G$. We see from eqn (5.6) that pairing correlations lead to a substantial reduction in the moment of inertia, and from eqn (5.17) that the coriolis force will have the effect of reducing the pairing correlations and hence the moment of inertia will rise towards the cranking-formula value.

Now let us consider the effect of the centrifugal force. At first sight the effect

† Both contribute to the energy in order ω^2 since the Hamiltonian must be time-reversal invariant.

is simply to add the potential (5.12) to the mean field. If we assume this to be of a Nilsson form then we have

$$V = V_{\mathrm{N}} + V_{\mathrm{cent}} = \tfrac{1}{2}m(\omega_x{}^2 x^2 + \omega_y{}^2 y^2 + \omega_z{}^2 z^2) - \tfrac{1}{2}m\omega^2(x^2 + y^2), \tag{5.19}$$

which can be written

$$V = \tfrac{1}{2}m(\omega_{x'}{}^2 x^2 + \omega_{y'}{}^2 y^2 + \omega_{z'}{}^2 z^2), \tag{5.20}$$

i.e. all we have done is change the deformation of the field. Indeed this is simply the centrifugal stretching of a rotating body. We might expect that if we increase the deformation of the nucleus the moment of inertia will increase and supplement the effect of the coriolis force. In fact the situation is much more complex, because as we increase the deformation we change the density of states at the Fermi surface (see Fig. 3.18, (pp. 50 and 51),and the strength of the pairing

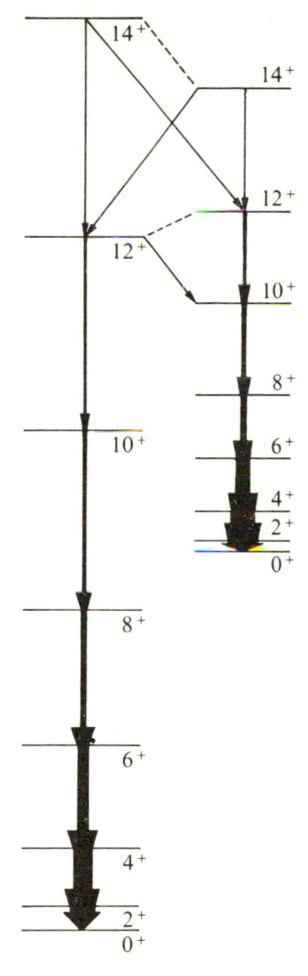

FIG. 5.5 Schematic example of band-crossing.

correlations depends critically on the density of single-particle states (see eqn (4.78)). Thus the centrifugal force can lead either to an increase or a decrease in the moment of inertia depending on which of these two effects is dominant in a given nucleus at a given deformation.

We turn now to the question of the back-bending observed in the heavy rare-earth nuclei and illustrated in Figs. 5.3 and 5.4. Neither of the phenomena discussed above can, in themselves, give a satisfactory account of the variation in the moment of inertia observed over the back-bending region. The simplest explanation of this effect is that there are two rotational bands, the ground-state band and an excited band based upon a shape isomer. If the excited band corresponds to a larger moment of inertia than the ground-state band then it will be compressed relative to the ground-state band as illustrated in Fig. 5.5. In this model the back-bending region is simply where the transition occurs from one band to the next. If this picture is correct then we would expect to observe a marked reduction in the E2 transition rates as we enter the back-bending region and to increase again above the transition region, and indeed this is the case. The reduction in the transition rates comes about because the transition operator, being a one-body operator, cannot strongly connect widely different intrinsic configurations.

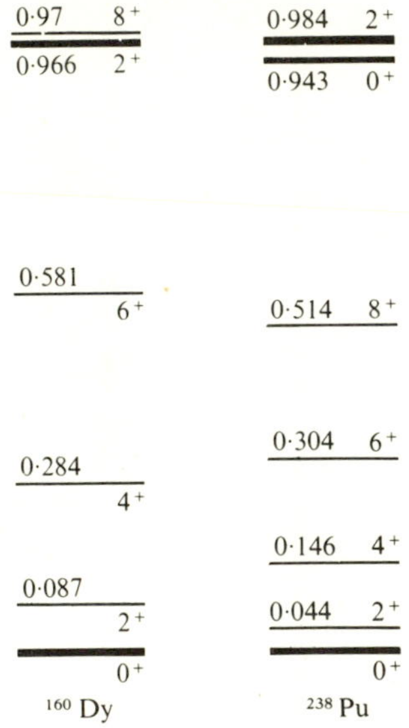

FIG. 5.6. A comparison of the spectra of a rare-earth nucleus ^{160}Dy with that of a heavy nucleus ^{238}Pu. This exhibits the similarity of pairing gap but the compression of the rotational band in the heavier nucleus.

There are two possible explanations of why the excited band should have a larger moment of inertia than the ground-state band: (1) the excited configuration may correspond simply to a much greater deformation; or (2) the pairing correlations may be substantially reduced. Since the residual interaction depends on the choice of mean field these two explanations may not be as different as they appear at first sight.

If we assume that the increased moment of inertia is due to a decreased pairing effect we are then faced with two schools of thought regarding the nature of the excited band. One argues that it is the reappearance of the intrinsic Hartree–Fock band and that it contains no pairing correlations. The other argues that since the coriolis force depends on the velocity of the particle (see eqns (5.11)) it will have a maximum effect in high spin states; it will break the pairing in

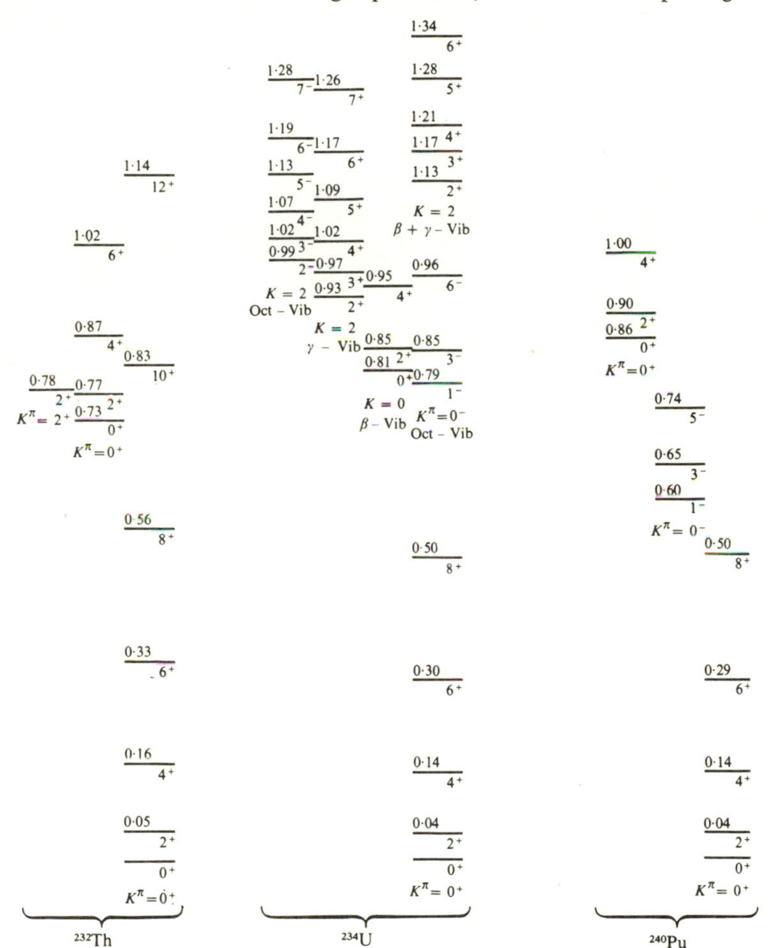

FIG. 5.7. Some examples of rotational bands in heavy nuclei.

these states first, creating a two-quasi-particle state and then, successively, the states of lower spin. Numerical calculations indicate that there is no great quantitative difference between these two models. This is perhaps not surprising since the dominant feature leading to a reduction of the moment of inertia is the pairing gap, and the major contribution to the pairing gap comes from the pairing in the highest spin states. At this level of theoretical analysis it is probably not sensible to argue about the distinction between the two views. Only a completely self-consistent calculation can tell in a particular case which of these extremely simplistic views is most appropriate.

To date, there is no evidence whatever of 'back-bending' in the heavy nuclei. This is not surprising since the moment of inertia of the heavy nuclei is typically about twice that of the rare earths. This follows immediately from the excitation energies of the first 2^+ state. In the rare earths this is typically $\sim 0{\cdot}1$ MeV, while in the heavy nuclei it is $\sim 0{\cdot}05$ MeV (this is illustrated with examples in Fig. 5.6). Thus the ground-state rotational band of the heavy nuclei is much compressed compared with that in the rare earths. However, the pairing gaps vary as $\sim A^{-\frac{1}{2}}$ and hence do not change greatly (see Fig. 5.6) at ~ 1 MeV. Thus the point at which an excited band catches up with the ground-state band, in the sense of Fig. 5.5, will occur at a much higher spin value in the heavy nuclei, and we have already commented on the difficulty of producing very high spin states in the heavy nuclei. Even if two bands do overlap this alone does not guarantee 'back-bending' since it is necessary to have substantial mixing of the bands in the critical region and this depends, amongst other things, on the levels in the two bands having relatively similar energies. As we go to high J the density of levels in a band is decreasing very rapidly and likewise the chance of levels from two bands having a similar energy is falling. Only very detailed calculations could predict whether 'back-bending' would or would not occur in a given nucleus.

We conclude this chapter with Fig. 5.7, in which are illustrated some examples of rotational bands so far identified in some heavy nuclei.

6

NUCLEAR STABILITY AND SHELL EFFECTS

In the last three chapters we have concentrated on the excitation spectra of the heavy nuclei, i.e. energies of states relative to the ground-state energy. If we wish to discuss the stability of the nucleus we require to know the *absolute* ground-state energy or at least its energy relative to the binding energy of the decay products. In the case of beta decay, and to a lesser extent alpha decay, it is sufficient to know only of the relative energies of a few neighbouring nuclei. However, if we are to discuss fission then we clearly need to know the ground-state energies of a large number of nuclei throughout the periodic table (Brack, Damgaard, Jensen, Pauli, Strutinsky, and Wong 1972; Strutinsky 1968; Wheeler 1955).

As an example of the task which we have set ourselves let us consider the case of ^{209}Bi already mentioned in Chapter 2, where we observed that we might expect ^{209}Bi to be unstable against alpha decay, i.e. the decay

$$^{209}\text{Bi} \rightarrow \alpha + {}^{205}\text{Tl} \tag{6.1}$$

should be possible. This is based on the observation that the binding energies of ^{209}Bi, ^{4}He, and ^{205}Tl are 1640·295 MeV, 28·297 MeV, and 1615·128 MeV respectively, thus making ^{209}Bi 3·13 MeV unstable against alpha decay, which should be compared with the Coulomb barrier of ~ 15 MeV for an alpha particle at the surface of a ^{205}Tl nucleus (see Fig. 6.1.). The fact that the alpha decay of

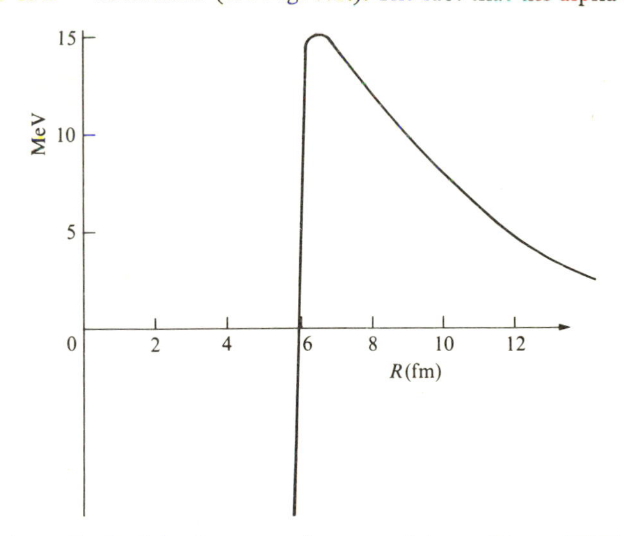

FIG. 6.1. Coulomb barrier appropriate to an alpha particle on. ^{205}Tl.

^{209}Bi is not observed must have to do with the penetrability of the Coulomb barrier by the alpha particle. To make sensible statements about this we require to map the alpha-particle—^{205}Tl interaction potential surface to an accuracy of at least 10 per cent, i.e. ~ 1 MeV. Similar considerations apply to fission, where the fission barrier of Fig. 3.12 (p. 39) must be calculated to around 1 MeV accuracy. If we try to calculate total binding energies then in the case of the heavy nuclei we are talking about ~ 1 MeV in a total binding energy of ~ 1·8 GeV, i.e. an accuracy of 0·06 per cent. There is no way that any present theory of nuclear structure can hope to calculate absolute energies to anything approaching this level of accuracy. Hence our approach to the problem must be a mixture of simple phenomenology coupled to plausible microscopic models.

We could fall back on the liquid-drop semi-empirical mass formula. But this only gives us the systematic energies in equilibrium configurations, and the variations of the liquid-drop parameters, e.g. surface area and Coulomb energy, are not fine enough to provide us with an accurate map of decay barriers. As we have seen in section 3.2 we require to supplement the liquid-drop energies with a pairing contribution and shell correlations. In Chapter 4 we began a study of the microscopic theory of the pairing correlations, and in this chapter we shall study the source of the shell correlations. The problem then becomes the coupling of the liquid-drop prescription for calculating binding energies, interpreted as giving the gross systematic variation of energies with a microscopic description of the detailed fluctuations to be expected in any particular nuclear configuration. We shall see that this combined approach is capable of explaining the observed binding energies of the heavy nuclei to ~ 1 MeV.

It will become obvious that the dominant fluctuations result from the variations in the density of single-particle states near the Fermi level. We have already seen the significance of this in the case of pairing correlations.

In a spherical well there are degeneracies of single-particle levels associated with the symmetries of the well, e.g. a single-particle level of angular momentum j is at least $(2j + 1)$ degenerate, and we refer to this unevenness in the distribution of single-particle levels as the shell structure. Thus in a shell there is a high density of state associated with the degeneracies, and the end of the shell is marked by a sudden fall in the density of states. We shall now generalize the concept of a shell and consider 'shell' closure to occur whenever there is such a dramatic fall in the density of states. This is schematically illustrated in Fig. 6.2, and in the Nilsson model it can occur 'accidentally' (examples may be found in Fig. 3.18, (pp. 50 and 51)) at various deformations.

Such shell closure corresponds to an increase in stability in a non-interacting system (i.e. an extreme single-particle model), since in order to produce a low density of states immediately outside a closed shell all the single-particle levels inside the shell must be depressed (see Fig. 6.3). This is not strictly true since the same effect could be achieved by raising the levels immediately above the Fermi level. While this would have no effect on the energy of the closed shell its stability

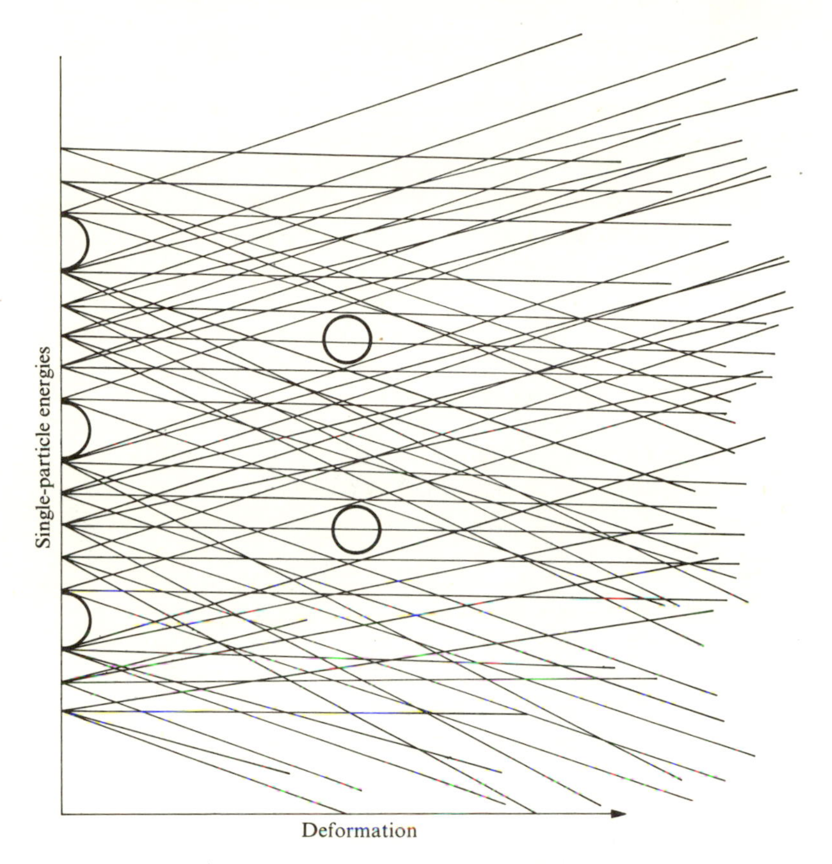

FIG. 6.2. Variation in density of single-particle states as a function of deformation. Circles enclose regions of low density corresponding to 'shells'.

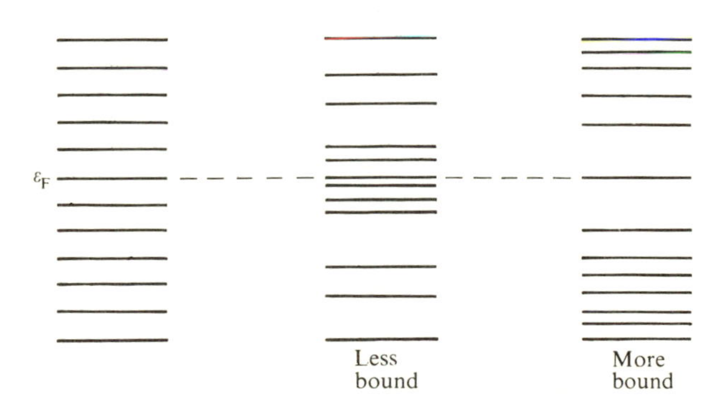

FIG. 6.3. Variations in density of single-particle states at the Fermi surface.

vis à vis the 'closed shell plus one particle' nucleus would again be increased. In an interacting system the situation is not quite so clear cut, since the correlation energy produced by the residual interaction may depend on the density of states at the Fermi surface, e.g. pairing correlations. And whether a nucleus is more or less stable as a result of inhomogenieties in the density of single-particle states will depend both on the shell effects and the correlation energy. In nuclei we know that as far as spherical shells are concerned the shell effect is stronger than the pairing effect, since the magic-number nuclei are so clearly associated with shell closure.

Thus our model for describing the ground-state energies of nuclei will be one in which the smooth gross variations are given by the liquid-drop model supplemented by shell effects and pairing energies,

$$E = E_{\mathrm{ld}} + \Delta E_{\mathrm{sh}} + \Delta E_{\mathrm{pair}}. \tag{6.3}$$

The liquid-drop energy will be assumed to be of the form

$$E_{\mathrm{ld}} = E_0(A, Z) + \Delta E_{\mathrm{def}}. \tag{6.4}$$

$E_0(A, Z)$ will be taken to be the energy predicted by the semi-empirical mass formula for spherical nuclei (see section 3.2) neglecting the shell and pairing terms. The deformation energy ΔE_{def} will measure the change in the liquid-drop energy from Coulomb (C) and surface (s) effects due to some distortion of the drop, as measured by some set of parameters β,

$$\Delta E_{\mathrm{def}}(\beta) = \Delta E_{\mathrm{C}}(\beta) + \Delta E_{\mathrm{s}}(\beta). \tag{6.5}$$

The deformations must be constrained such that the volume of the drop is constant, this being consistent with the observed high incompressibility of nuclear matter.

In order to calculate the shell and pairing terms in eqn (6.3) we shall assume that there is a mean field in which the nucleons move. For self-consistency we shall assume that this potential has the same shape as the liquid drop, and hence the single-particle wavefunctions, energies, and density of states will all be functions of the deformation parameters β. The pairing energy will be calculated exactly as in section 4.3 using the single-particle states of this distorted well. The shell corrections will arise from fluctuations in the density of single-particle states described by the parameters β, as discussed below. In Fig. 6.4 we illustrate this variation in density of single-particle states in the case of spherical Wood–Saxon well and a prolate spheroid of the same Wood–Saxon form and volume,

$$U_{WS} = V_0\{1 - \exp(r - R/d)\}^{-1}. \tag{6.6}$$

The distortion parameter in this case is taken to be the eccentricity of the spheroid ϵ. In the spherical case $\epsilon = 0$ the correlation between low density of states and the well-known magic numbers is evident. We draw attention to the fact that Fig. 6.4 shows that the next doubly magic nucleus after ^{208}Pb, with $N = 126$ and $Z = 82$, should occur for $N \sim 184$ and $Z \sim 112$ or 114. We shall return to this point in Chapter 8. In the case of a deformation $\epsilon = 0{\cdot}3$ we see an almost complete anticorrelation between the low density of states and the spherical magic numbers. Indeed at this deformation there seems to be increase stability for nuclei in the group ^{236}U, ^{238}U, ^{240}Pu.

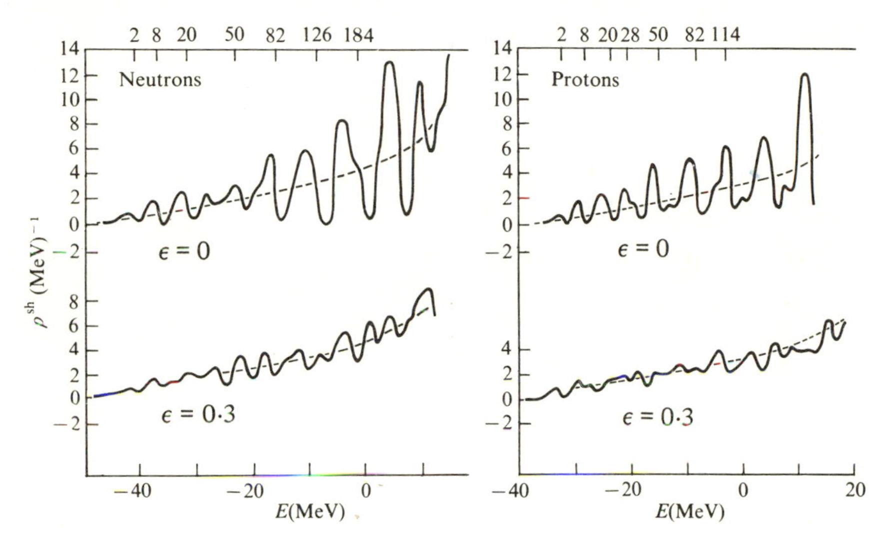

FIG. 6.4. Variations in density of single-particle states in the Wood–Saxon well of eqn (6.6) as a function of single-particle energy for zero distortion $\epsilon = 0$ and a spheroid with eccentricity $\epsilon = 0{\cdot}3$. The shell density ρ^{sh} is defined by eqns (6.20) and (6.26).

In Fig. 6.4 we have plotted a smooth curve representing the density of single-particle levels, while in practice, if we solve a given Schrödinger equation, we shall obtain, at least for the bound states, a set of perfectly discrete levels ϵ_1, ϵ_2 . . . of varying degeneracies g_1, g_2 . . . . The density of levels is then

$$\rho(E) = \sum_i g_i \, \delta(E - \epsilon_i), \tag{6.7}$$

and the total non-interacting energy of the system will be

$$E_0 = \int_{-\infty}^{\infty} E\rho(E)\mathrm{d}E = \sum_{i=1}^{F} g_i \, \epsilon_i. \tag{6.8}$$

It is anticipated that the bulk of this energy corresponds to E_{ld} of eqn (6.4), and in calculating ΔE_{sh} we are interested in calculating the variation about some mean

E_0 caused by abnormally high or abnormally low densities at the Fermi level. We have thus set ourselves three problems:

(1) how to define a smooth *average* density $\bar{\rho}$,

(2) how to define a smooth fluctuating density ρ^{sh},

(3) how to define the Fermi level.

Let us consider the last problem first. The Fermi energy is defined by

$$N = \int_{-\infty}^{\epsilon_{\text{F}}} \rho(E)\,\mathrm{d}E, \tag{6.9}$$

where N is the number of particles in the system. In the nucleus we will treat the neutron and protons separately, but for the moment we will restrict ourselves to one kind of particle. Now if we have two versions of the density of levels we will obtain two Fermi energies $\bar{\epsilon}_{\text{F}}$ and $\epsilon_{\text{F}}{}^{\text{sh}}$ defined by

$$N = \int_{-\infty}^{\epsilon_{\text{F}}} \bar{\rho}(E)\,\mathrm{d}E,$$

$$N = \int_{-\infty}^{\epsilon_{\text{F}}^{\text{sh}}} \rho^{\text{sh}}(E)\,\mathrm{d}E. \tag{6.10}$$

Normally the difference between these definitions will be required to be small in the sense that

$$\epsilon_{\text{F}}^{\text{sh}} - \bar{\epsilon}_{\text{F}} \leqslant 1/\bar{\rho}\,(\bar{\epsilon}_{\text{F}}) \ll \bar{\epsilon}_{\text{F}}, \tag{6.11}$$

while the shell correction energy we shall define as

$$\Delta E_{\text{sh}} = \int_{-\infty}^{\epsilon_{\text{F}}^{\text{sh}}} E\rho^{\text{sh}}(E)\,\mathrm{d}E - \int_{-\infty}^{\bar{\epsilon}_{\text{F}}} E\bar{\rho}(E)\,\mathrm{d}E. \tag{6.12}$$

Combining eqns (6.10) and (6.11) we have

$$\epsilon_{\text{F}}^{\text{sh}} - \bar{\epsilon}_{\text{F}} = -\int_{-\infty}^{\epsilon_{\text{F}}^{\text{sh}}} (\rho^{\text{sh}}(E) - \bar{\rho}(E))\,\mathrm{d}E \,/\, \bar{\rho}(\epsilon_{\text{F}}^{\text{sh}}), \tag{6.13}$$

and inserting this in eqn (6.12) and expanding in a Taylor series about $\epsilon_{\text{F}}^{\text{sh}}$ we have

$$\Delta E_{\text{sh}} = \int_{-\infty}^{\epsilon_{\text{F}}^{\text{sh}}} (E - \epsilon_{\text{F}}^{\text{sh}})\left\{\rho^{\text{sh}}(E) - \bar{\rho}(E)\right\}\mathrm{d}E - O\left\{(\epsilon_{\text{F}}^{\text{sh}} - \bar{\epsilon}_{\text{F}})^2\right\}. \tag{6.14}$$

Keeping to lowest order and from now on suppressing the distinction between ϵ_F^{sh} and $\bar{\epsilon}_F$ we have

$$\Delta E_{sh} = \int_{-\infty}^{\epsilon_F} (E - \epsilon_F)\{\rho^{sh}(E) - \bar{\rho}(E)\}\,dE, \tag{6.15}$$

and in actual calculations ϵ_F will be chosen to have a value halfway between the last occupied and the first empty level in the mean field. From eqn (6.15) it is clear that

$$\begin{aligned}\Delta E_{sh} &> 0, \quad \rho^{sh}(\epsilon_F) < \bar{\rho}(\epsilon_F)\\ &< 0, \quad \rho^{sh}(\epsilon_F) > \bar{\rho}(\epsilon_F),\end{aligned} \tag{6.16}$$

as required.

In order to define smooth density functions ρ^{sh} and $\bar{\rho}$ we average over the extreme shell model levels (6.7) thus

$$\bar{\rho}(E) = \bar{\gamma}^{-1}\int_{-\infty}^{\infty} \xi\{(E-E')/\bar{\gamma}\}\rho(E')\,dE' = \bar{\gamma}^{-1}\sum_i \xi\{(E-E_i)/\bar{\gamma}\}. \tag{6.17}$$

Clearly, there is no unique prescription for carrying our this smoothing prescription. We shall require two energy ranges $\bar{\gamma}$ and γ^{sh} over which we average in order to obtain the two smooth density functions $\bar{\rho}$ and ρ^{sh}. In order that $\bar{\rho}$ does not show shell effects $\bar{\gamma}$ must be greater than or equal to the spacing between shells, while in order that ρ^{sh} shows the shell effects and yet is smooth we require that γ^{sh} be large compared with spacing between levels within a shell but small compared with the spacing between shells. Two constraints must be applied to the smoothing function ξ – these are

$$\int_{-\infty}^{\epsilon_F} \bar{\rho}(E)\,dE = N \tag{6.18}$$

and

$$\bar{\rho}(E) \underset{\bar{\gamma}\to 0}{\to} \rho(E). \tag{6.19}$$

The simplest smoothing is via a simple Gaussian function whence

$$\bar{\rho}(E) = (\pi^{\frac{1}{2}}\bar{\gamma})^{-1}\sum_i \exp\left[-\{(E-E_i)/\bar{\gamma}\}^2\right]. \tag{6.20}$$

If we assume that the mean field is of oscillator form then the spacing between major shells in $\hbar\omega_0$. If we then calculate the root-mean square radius of the charge density and fit it to the observed radii $R = r_0 A^{\frac{1}{3}}$ ($r_0 \sim 1{\cdot}2$ fm) then we find empirically that

$$\hbar\omega_0 \simeq 41 A^{-\frac{1}{3}} \sim 7\text{–}10 \text{ MeV}, \tag{6.21}$$

or more generally Bohr's quantization rule for a particle in a potential of finite size R is

$$p_n R \simeq n\hbar \tag{6.22}$$

where p_n is the momentum of particle with energy ϵ_n. Thus the spacing between levels is

$$\hbar\omega_0 = \epsilon_{n+1} - \epsilon_n \simeq \frac{\hbar}{R}\left(\frac{2\epsilon_n}{m}\right)^{\frac{1}{2}} \tag{6.23}$$

at the Fermi surface $\epsilon_n = \epsilon_F$. Using a Fermi-gas assumption

$$\epsilon_F = \frac{\hbar^2}{2m}\left(\frac{9}{64\pi^2}\,\frac{1}{r_0{}^3}\right)^{\frac{2}{3}}, \tag{6.24}$$

we have

$$\hbar\omega_0 \simeq 5\text{–}10 \text{ MeV}, \tag{6.25}$$

in agreement with eqn (6.21). For the smooth average density we shall use $\bar{\gamma} = \hbar\omega_0$. The smoothing width γ^{sh} must be large enough to include many levels within a shell but not large enough to produce any interference between shells; suitable values are

$$\gamma^{sh} \simeq \hbar\omega_0 / A^{\frac{1}{3}} \simeq 1\text{–}2 \text{ MeV}. \tag{6.26}$$

We are now in a position to calculate the total binding energy of the nucleus as a function of the deformation parameters β and the mass and charge numbers A and Z. The simplest calculation involves a Nilsson-type mean field and a corresponding quadrupolely deformed liquid drop. Fig. 6.5 shows a comparison of such a calculation with the observed nuclear masses, and we see that typically the discrepancy $\lesssim 1$ MeV. More detailed calculations have fitted more than 1000 nuclei with a root-mean-square deviation of $\lesssim 0{\cdot}65$ MeV.

The equilibrium deformations are determined by the equations

$$\partial E(\beta) / \partial\beta = 0, \tag{6.27}$$

and in general there may be more than one solution to eqn (6.27) corresponding to the existence of several stable, or metastable, shape isomers. Of particular interest is the existence of fission isomers to be discussed in Chapter 7.

Since 'shells' as we have defined them can occur at various deformations it is of interest to ask why the observed magic numbers all correspond to spherical shells. The reason is that the shell effect is maximized when the fall in the density of single-particle levels is maximized. Now, apart from 'accidental' degeneracies, degeneracy of a level implies some symmetry of the potential well, and in three dimensions the figure of maximum symmetry is the sphere. Hence the degeneracy of single-particle levels is maximized in a spherical potential and the shell effect is most obvious.

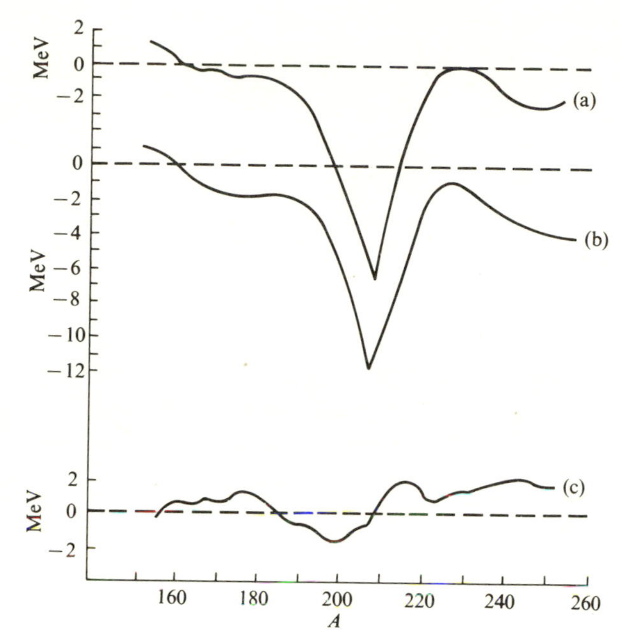

FIG. 6.5. A comparison between observed and calculated nuclear masses. (a) Experimental mass – liquid-drop mass; (b) Shell correction energies; (c) Discrepancy between calculated masses and experiment.

The greatest remaining uncertainty surrounds the treatment of the pairing correlations. Here we shall briefly discuss two problems.

(1) Should the pairing correction in eqn (6.3) be the total pairing energy, as we have suggested, or should it simply be a measure of the fluctuation in the pairing energy about some smoothed contribution which is already represented in either E_{ld} or ΔE_{sh}?

(2) The pairing energy depends on the deformation through the density of single-particle levels, but is the pairing strength G itself a function of deformation? One would expect it to be so, since surely the matrix element depends on the single-particle states which are functions of the deformation.

The first problem is easily taken into account by defining

$$\Delta E_{\mathrm{pair}} = E_{\mathrm{pair}} - \bar{E}_{\mathrm{pair}} \tag{6.28}$$

where, according to eqn (4.80), we have

$$\bar{E}_{\mathrm{pair}} = -\tfrac{1}{2}\,\bar{\rho}(\beta)\bar{\Delta}^2, \tag{6.29}$$

and we use the empirical gap parameter

$$\bar{\Delta} = 12/A^{\frac{1}{2}} \text{ MeV} \tag{6.30}$$

from section 3.2.1. The pairing energy is given by eqn (4.79),

$$E_{\text{pair}} = \sum_i [(\epsilon_i - \epsilon_F) - \{(\epsilon_i - \epsilon_F)^2 + \tfrac{1}{2}\Delta^2\} \;/\; \{(\epsilon_i - \epsilon_F)^2\}^{\frac{1}{2}}]. \tag{6.31}$$

In connection with the latter problem, it has been suggested that because of the reduced effect of the exclusion principle the residual nucleon–nucleon interaction is most attractive in the surface of the nucleus and hence the pairing strength will be greatest in the surface. Some calculations have been carried out in the so-called surface-pairing approximation, in which the pairing strength is assumed proportional to the surface area. This is equivalent to the assumption that the surface tension is not constant but increases as the surface is stretched.

Only a detailed self-consistent calculation would give definitive answers to the above questions. However, two general points may be made: the pairing effect may be expected to smooth out the shell effects to some extent, and the pairing gap generally increases with increasing distortion of the nucleus.

7

FISSION AND ALPHA DECAY

So far we have devoted a considerable amount of attention to the study of deformations of nuclear shapes. In the case of ground-state rotational and vibrational bands the departures from sphericity are naturally quite small, characterized by Nilsson parameters $\delta \lesssim 0{\cdot}3$ (eqn 3.81), p. 50). The processes of fission, and to a lesser extent alpha decay, may be viewed as a study of the deformations of nuclear shapes carried to the extreme limit, resulting in the nucleus separating into two or more fragments. We shall only discuss binary fission, although it will be clear that similar techniques could be applied to the study of ternary and higher-order fission with a corresponding increase in the computational complexity.

We shall again make the assumption of adiabaticity, i.e. the velocity of collective motion, in this case the separation of the fragments, is assumed much slower than the motions of individual nucleons within the nucleus. While this assumption is probably reasonable during the critical stage in the formation of the fragments, it is probably not valid in the latter stages of the process leading to fission (Wilets 1964).

In section 3.2.4 we have already introduced the basic features used in the liquid-drop model to describe the fission process. We shall now develop these ideas in greater detail and show that they are insufficient in many respects to give a satisfactory explanation of the observed features of fission. We shall then show how these deficiencies may be rectified by including some aspects of the microscopic motions of the nucleons.

We shall begin by using eqn (3.43) to describe the surface of the nuclear drop

$$R(\theta) = R_0\left(1 + \sum_{n=z}^{N} \alpha_n P_n(\cos\theta)\right), \tag{7.1}$$

where the radius of the undeformed spherical drop is

$$R_0 = 1{\cdot}2\, A^{\frac{1}{3}}\ \mathrm{fm}. \tag{7.2}$$

The coefficients α_n are generalized deformation coordinates constrained by the condition of nuclear incompressibility.

$$\int_0^{\pi} \pi R^2(\theta)\left(-\frac{\mathrm{d}R(\theta)}{\mathrm{d}\theta}\cos\theta + R\sin\theta\right)\sin^2\theta\ \mathrm{d}\theta = \tfrac{4}{3}\pi R_0^3 = 7{\cdot}25\,A\ \mathrm{fm}^3, \tag{7.3}$$

and the constancy of the centre-of-mass coordinate (chosen to be the origin),

$$\int_0^\pi R^3(\theta)\left(-\frac{dR(\theta)}{d\theta}\cos\theta + R\sin\theta\right)\cos\theta\sin^2\theta\, d\theta = 0. \quad (7.4)$$

Note that we excluded P_0 and P_1 terms from eqn (7.1), as explained in Chapter 3 (p. 29). Had they been included eqn (7.3) would have required $\alpha_0 = 0$ while eqn (7.4) would require $\alpha_1 = 0$. It should be further noted that the centre of mass does not necessarily coincide with the centre of the neck during an asymmetric fission process.

The surface energy is proportional to the surface area and for a spherical drop is simply

$$E_S(0) = 4\pi R_0^2 S, \quad (7.5)$$

where S is the surface tension. The surface energy of the deformed drop is

$$E_S(\alpha_2, \alpha_3, \ldots) = 4\pi S \int_0^\pi R^2(\theta)\left(1 + \frac{1}{R(\theta)}\frac{dR(\theta)}{d\theta}\right)^{\frac{1}{2}} \sin\theta\, d\theta, \quad (7.6)$$

which, if the deformation parameters are small, may be written

$$E_S(\alpha_2, \alpha_3, \ldots) = 4\pi R_0^2 S\left(1 + \frac{2}{5}\alpha_2^2 + \frac{5}{7}\alpha_3^2 + \ldots\right). \quad (7.7)$$

The Coulomb energy of a charge distribution of density $\rho(r)$ is

$$E_C = \tfrac{1}{2}\int \frac{\rho(\mathbf{r})\,\rho(\mathbf{r}')}{|\mathbf{r}-\mathbf{r}'|}\, d\mathbf{r}\, d\mathbf{r}'. \quad (7.8)$$

Assuming that the charge is uniformly distributed throughout the nuclear drop and that the volume remains constant we obtain a constant charge density

$$\rho = Ze/\tfrac{4}{3}\pi R_0^3 = 0{\cdot}414\, Ze/A \quad (7.9)$$

For a spherical distribution eqn (7.8) yields

$$E_C(0) = \tfrac{3}{5}(Ze)^2/1{\cdot}2A^{\frac{1}{3}}, \quad (7.10)$$

and for small distortions we obtain

$$E_C(\alpha_2, \alpha_3, \ldots) = \frac{3}{5}\frac{(Ze)^2}{1{\cdot}2A^{\frac{1}{3}}}\left(1 - \frac{1}{5}\alpha_2^2 - \frac{10}{49}\alpha_3^2 \ldots\right). \quad (7.11)$$

The deformation energy is obtained by combining eqns (7.7) and (7.11),

$$\Delta E = \left(\frac{2}{5}\alpha_2^2 + \frac{5}{7}\alpha_3^2 + \ldots\right)E_S(0) + \left(-\frac{1}{5}\alpha_2^2 - \frac{10}{49}\alpha_3^2 \ldots\right)E_C(0). \quad (7.12)$$

Restricting ourselves to quadrupole deformations we have

$$\Delta E \simeq \frac{2}{5} \alpha_2^2 \; E_s(0)\,(1-\chi), \tag{7.13}$$

where χ is the fissility parameter

$$\chi = \frac{Z^2}{A}\left(\frac{3e^2}{40\pi\,(1{\cdot}2)^3 S}\right). \tag{7.14}$$

A simple interpretation of eqn (7.13) would be that for $\chi < 1$ the drop is stable against spontaneous fission and when $\chi > 1$ the nucleus is unstable against spontaneous fission. The critical case $\chi = 1$ corresponds to the critical ratio of eqns (3.24) and (3.42)

$$\left(\frac{Z^2}{A}\right)_{\text{crit}} = \frac{40\pi(1{\cdot}2)^3 S}{3e^2} \simeq 40\text{–}50. \tag{7.15}$$

In Fig. 7.1 we again reproduce the plot of spontaneous fission half-lives versus

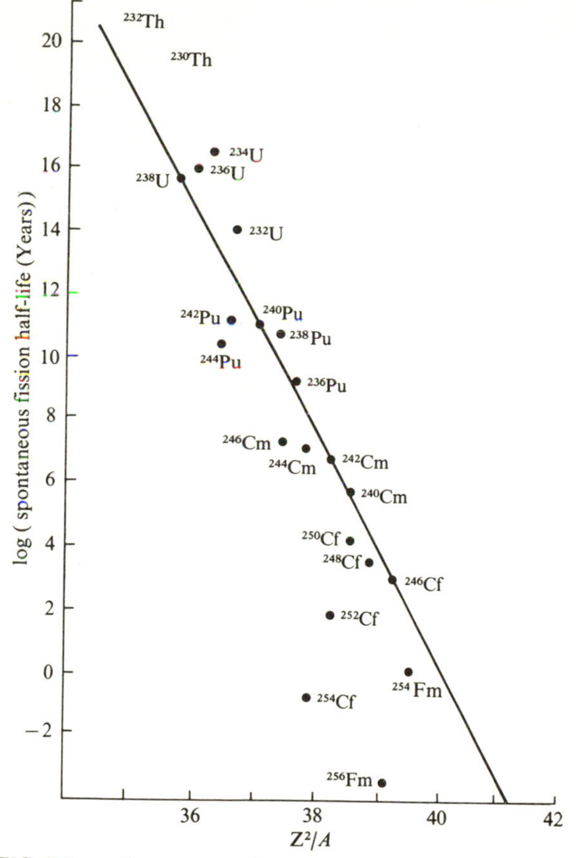

FIG. 7.1. Spontaneous fission half-lives versus Z^2/A.

Z^2/A, and we see that there is an approximately straight line envelope intersecting the axis at $Z^2/A \simeq 45$. However, eqn (7.13) is only valid for small α; hence it only sets limits on the stability against distortion and does not describe the situation in the case of large deformations near the scission point. There is nothing in the liquid-drop model to shed light on the different spontaneous fission properties of even and odd isotopes.

The liquid-drop model also fails to explain the observed fission yields. To see this consider the simple model of Fig. 7.2 in which an originally spherical nucleus of mass and charge number A_0 and Z_0 has fissioned into two identical spherical nuclei which are just touching. The total potential energy in this configuration is

$$E_t(0) = 2\left\{4\pi R^2 S + \frac{3}{5}(Ze)^2/R\right\} + (Ze)^2/2R, \tag{7.16}$$

where $Z = \frac{1}{2}Z_0$, $A = \frac{1}{2}A_0$, and

$$R = r_0 A^{\frac{1}{3}}. \tag{7.17}$$

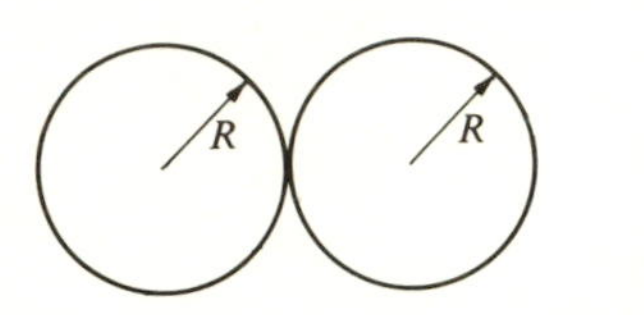

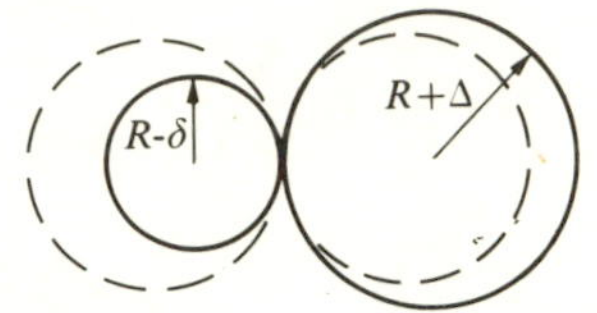

FIG. 7.2. Asymmetry in fission

Now consider an infinitesimal change from this symmetric case in which the left-hand sphere loses nuclear matter to the right-hand sphere such that the radius of the left-hand sphere is reduced to $(R - \delta)$ while the radius of the right-hand sphere increases to $(R + \Delta)$. Since the total volume is conserved, we have

$$(R + \Delta)^3 + (R - \delta)^3 = 2R^3, \tag{7.18}$$

whence, to second order,

$$\delta = \Delta(1 + 2\Delta/R). \tag{7.19}$$

Defining $\xi = \Delta/R$ the total energy now becomes

$$E_t(\xi) = E_t(0) + \xi^2\left(-2E_s(0) + \frac{10}{3}E_C(0)\right). \tag{7.20}$$

Hence the change in energy is

$$\Delta E_t = 2E_s(0)\,\xi^2\left(\frac{10}{3}\chi - 1\right). \tag{7.21}$$

The fissility parameter χ_0 of the parent nucleus is

$$\chi_0 \simeq \frac{Z_0^2}{A_0}\Big/50 = \frac{(2Z)^2}{2A}\Big/50 = 2\chi, \tag{7.22}$$

and the change in energy may be written

$$\Delta E_t = 2E_s(0)\,\xi^2\left(\frac{5}{3}\chi - 1\right). \tag{7.23}$$

We see that

$$\Delta E_t > 0, \quad \chi_0 > 0{\cdot}6, \quad \text{i.e. } A \gtrsim 200 \tag{7.24}$$

and

$$\Delta E_t < 0, \quad \chi_0 < 0{\cdot}6, \quad \text{i.e. } A \lesssim 200, \tag{7.25}$$

corresponding to symmetric fission being favoured in very heavy nuclei and asymmetric fission being favoured in lighter nuclei. This is contrary to observation in which most of the heavy fissioning nuclei exhibit asymmetric yields.

We shall now turn our back on eqn (7.1) since a single-centred form seems inappropriate for the discussion of a system which is going to separate into two fragments. Of course, there is a vast choice of two-centred forms for the nuclear drop, but we shall restrict our discussion to the form illustrated in Fig. 7.3,

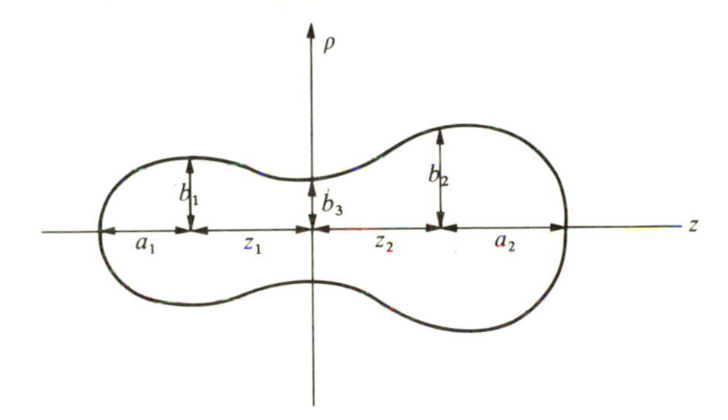

FIG. 7.3. Two-centre model of fission.

which consists of two intersecting spheroids with major axes a_1 and a_2 and minor axes b_1 and b_2. In order to avoid any unphysical discontinuity in the slope of the surface the region $\frac{1}{2}z_1 < z < \frac{1}{2}z_2$ are joined by a hyperboloid. Thus in cylindrical polar coordinates the surface is given by

$$\begin{aligned}
\frac{\rho^2}{b_1^2} &= 1 - \frac{(z-z_1)^2}{a_1^2}, \quad z < -\tfrac{1}{2}z_1, \\
\frac{\rho^2}{b_3^2} &= 1 + \frac{z^2}{a_3^2}, \quad -\tfrac{1}{2}z_1 \leqslant z \leqslant \tfrac{1}{2}z_2, \\
\frac{\rho^2}{b_2^2} &= 1 - \frac{(z-z_2)^2}{a_2^2}, \quad z > \tfrac{1}{2}z_2.
\end{aligned} \tag{7.26}$$

If the conditions $b_1 = b_2 = b$ and $z_1/a_1 = z_2/a_2$ are imposed (they are unnecessary, but will simplify our analysis later) then the continuity of the surface and its derivatives at $z = \frac{1}{2}z_1$ and $z = \frac{1}{2}z_2$ yield

$$a_3 = \begin{cases} a_2\,(1 - z_2^2/a_2^2)^{\frac{1}{2}}, & z > 0 \\ a_1\,(1 - z_1^2/a_1^2)^{\frac{1}{2}}, & z \leqslant 0. \end{cases} \tag{7.27}$$

Treating the nuclear drop as being incompressible the ratio of the masses in the two fragments is given by

$$A_2/A_1 = a_2/a_1, \tag{7.28}$$

and the volume is constrained to be constant and equal to

$$\frac{4}{3}\pi(A_1 + A_2)r_0^3. \tag{7.29}$$

The surface energy and Coulomb energy can now be calculated as a function of the three independent parameters b, $p = A_2/A_1$, and $z_0 = z_1 + z_2$. Fig. 7.4 shows a contour plot of the potential energy in MeV in the plane of b and z_0 for the

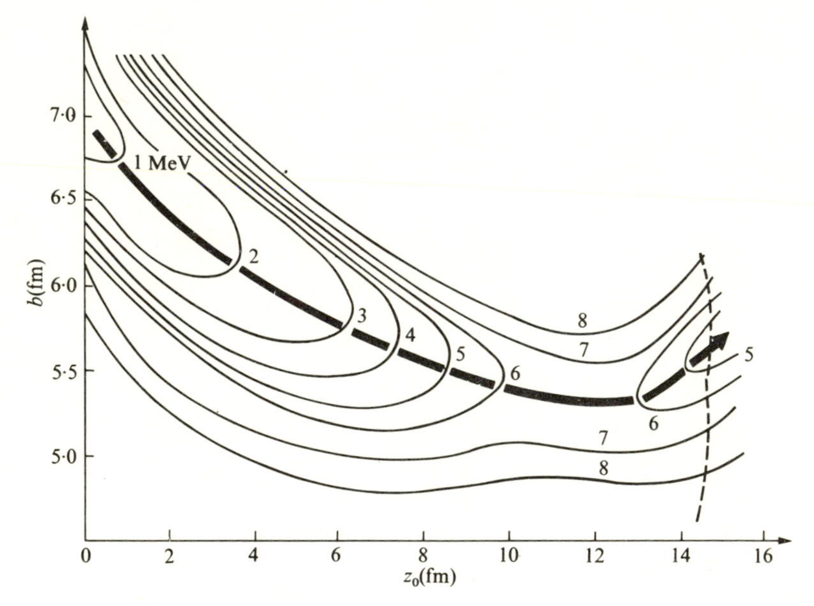

FIG. 7.4. Potential-energy contour plot (in MeV) for symmetric liquid-drop fission of ^{236}U in the plane of minor axis and centre separation. The heavy line indicates the fission path.

case of ^{236}U with $p = 1$. Consistent with our assumption of adiabaticity we shall assume that $\dot{z}_0$ is so slow that the nucleons can adjust their motion to optimize the values of b and p in order to follow the minimum potential-energy path in the countour plane of Fig. 7.4. In Fig. 7.5 the potential energy contours are plotted for ^{236}U in the plane of p and z_0 where at each point the value of b has been optimized. Consistent with our earlier analysis we see that the fission path crosses

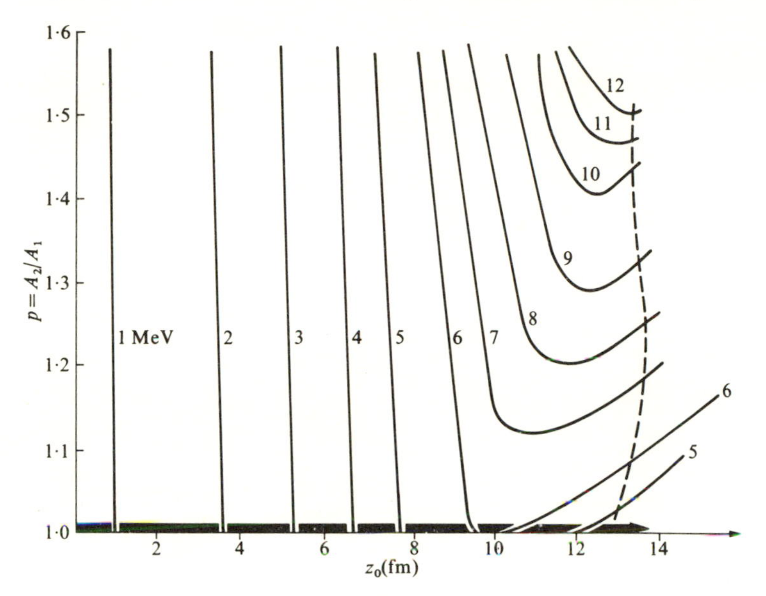

FIG. 7.5. As Fig. 7.4 in the plane of mass asymmetry and centre separation. The minor axis is that which corresponds at each separation to the fission path of Fig. 7.4.

the scission line at $p = 1$, i.e. symmetric fission. Fig. 7.6 shows the potential-energy cross-section along the fission path, and we see the existence of the fission barrier already suggested in section 3.2.4. The height of the barrier in Fig. 7.6

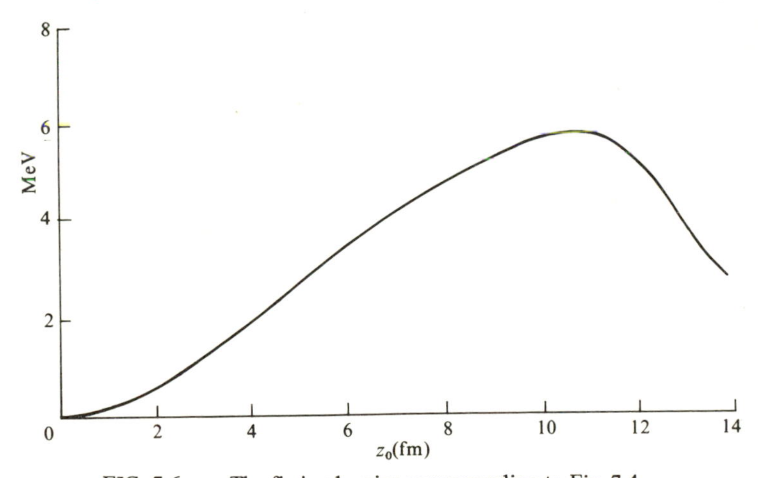

FIG. 7.6. The fission barrier corresponding to Fig. 7.4.

is ~ 8 MeV, and this should be compared with the observed fission thresholds of ~ 6 MeV. The height of the fission barrier is extremely sensitive to the smooth joining of the neck, in our case by a hyperboloid (eqn (7.26)). If the neck is not smoothed then the barrier rises to 15–20 MeV owing principally to the increased surface energy.

Assuming that the gross features of the fission process are given by the model outlined above we now seek to study the effects of shell corrections and pairing correlations described in Chapter 5.

Neglecting the neck region for the moment, the single particles find themselves moving in a double Nilsson well where the equipotential surfaces are continuous at $z = 0$ and the potential deformations are consistent with the liquid-drop deformations (Fig. 7.7). Neglecting the non-central terms, the Schrodinger equation

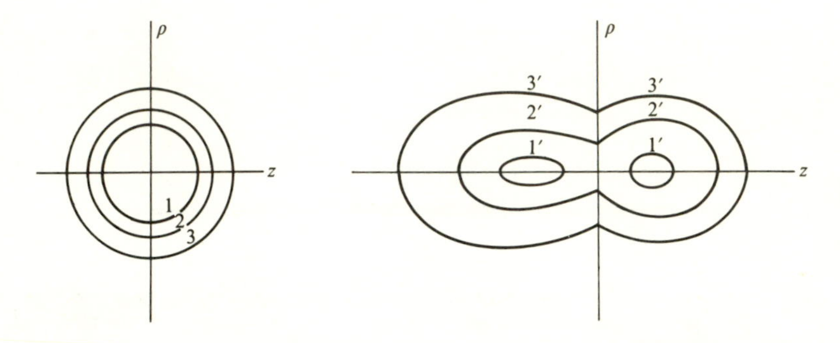

FIG. 7.7. Equipotentials in a spherical and two-centre shell model.

separates in cylindrical coordinates with single-particle states $\Psi(r) = Z(z)R(\rho)\Phi(\phi)$ (Holzer, Mosel, and Greiner 1967, 1971),

$$\left(\frac{\partial^2}{\partial\phi^2} + m_\phi^2\right)\Phi(\phi) = 0, \tag{7.30a}$$

$$\left\{\frac{\partial^2}{\partial\rho^2} + \frac{1}{\rho}\frac{\partial}{\partial\rho} - \frac{m_\phi^2}{\rho^2} - \frac{m^2\omega^2\rho^2}{\hbar^2} + \frac{4m\omega_\rho}{\hbar}\left(n_\rho - \frac{2}{1+|m_\phi|}\right)\right\}R(\rho) = 0, \tag{7.30b}$$

$$\left\{\frac{\partial^2}{\partial z^2} - \frac{m\omega_{z_1}^2}{\hbar^2}\left(z^2 + 2z_1 z + z_1^2 - \frac{2E}{m\omega_{z_1}^2}\right) + \frac{4m\omega_\rho}{\hbar}\left(n_\rho + \frac{2}{1+|m_\phi|}\right)\right\}Z(z) = 0,$$
$$z \leqslant 0$$

$$\left\{\frac{\partial^2}{\partial z^2} - \frac{m\omega_{z_2}^2}{\hbar^2}\left(z^2 - 2z_2 z + z_2^2 - \frac{2E}{m\omega_{z_1}^2}\right) + \frac{4m\omega_\rho}{\hbar}\left(n_\rho + \frac{2}{1+|m_\phi|}\right)\right\}Z(z) = 0,$$
$$z \geqslant 0. \tag{7.30c}$$

These equations have the solutions

$$\Phi(\phi) = \frac{1}{\sqrt{(2\pi)}} \exp(im_\phi \phi), \tag{7.31}$$

$$\chi(\rho) = \sqrt{\left\{\frac{2n_\rho}{(|m_\phi| + n_\rho)!}\right\}}\, k_\rho^{\frac{1}{2}(|m_\phi|+1)} (-1)^{\frac{1}{2}(|m_\phi| + n_\rho)} \exp(\tfrac{1}{2}k\rho\rho^2)\rho^{|m_\phi|} \times$$

$$\times\, L_{n_\rho}^{|m_\phi|}(k_\rho \rho^2) \tag{7.32}$$

and

$$Z(z) = \begin{cases} N_1 \exp\left\{-k_{z_1}(z - z_1)^2\right\} & U(-\tfrac{1}{2}n_{z_1}, \tfrac{1}{2}, k_{z_1}(-z - z_1)^2)\; z \leqslant 0 \\ N_2 \exp\left\{-k_{z_2}(z - z_2)^2\right\} & U(-\tfrac{1}{2}n_{z_2}, \tfrac{1}{2}, k_{z_2}(z - z_2)^2)\; z \geqslant 0, \end{cases} \tag{7.33}$$

where $m_\phi = 0_1 \pm 1, \pm 2, \pm 3$ represents the conserved component of angular momentum about the axis of symmetry, $k_\rho = m\omega\rho/\hbar$, with $n_\rho = 0, 1, 2, 3$, etc. corresponding to the usual oscillator quanta in the ρ degree of freedom (remember $b_1 = b_2 = b$). The functions U of eqn (7.33) are Kummel functions of the second kind and are related to the usual oscillator functions. The corresponding energies are

$$\epsilon_i = \hbar\omega_{zi}(n_{zi} + \tfrac{1}{2}), \quad i = 1, 2, \tag{7.34}$$

except that the n_{zi} are now no longer integers. The eigenvalues of n_{zi} are found by numerically matching the wavefunction Z and its derivative at $z = 0$. The constants N_i are then given by the usual normalization condition.

Having solved eqns (7.30) to obtain a set of single-particle basis functions the effect of the non-central terms in the Hamiltonian can be included by diagonalizing the matrix

$$\langle \Psi' | \mathcal{H} | \Psi \rangle = E\delta_{n_z n'_z}\delta_{n_\rho n_\rho'}\delta_{m_\phi m_\phi'}\delta_{ss'} + \langle \Psi' | \mathcal{H}_{nC} | \Psi \rangle, \tag{7.35}$$

where

$$\langle \Psi' | \mathcal{H}_{nC} | \Psi \rangle = -\kappa\hbar\omega_\rho (2\langle \Psi' | \mathbf{l}\,.\,\mathbf{s} | \Psi \rangle + \mu \langle \Psi' | \mathbf{l}^2 | \Psi \rangle), \tag{7.36}$$

Writing $\mathbf{l}\,.\,\mathbf{s} = \tfrac{1}{2}(l^+ s^- + l^- s^+) + l_z s_z$ (7.37)

and

$$\mathbf{l}^2 = \tfrac{1}{2}(l^+ l^- + l^- l^+) + l_z^2, \tag{7.38}$$

remembering that l_z is conserved and equal to $m_\phi \hbar$, simple selection rules can be found for the matrix elements (7.36). The only minor sophistication is that the strengths κ and μ in the Nilsson model are adjusted to reproduce the spin-orbit splittings in the spherical nuclei and these are functions of mass; typical values are given in Table 7.1. In fission we go from a heavy actinide to two rare-earth fragments hence we have used for κ and μ a linear interpolation between the two sets of values in Table 7.1, i.e. κ and μ are functions of z_0 and when $z_0 = 0$ they correspond to actinide values and when $z_0 = z_{\text{scission}}$ they take the rare-earth values.

Finally, there is the question of the neck. This represents a perturbation to the two intersecting Nilsson potentials (Fig. 7.8). The perturbation is of the form

$$V_{\text{neck}} = \begin{cases} -m\omega_{z_2}^2 (z - \tfrac{1}{2} z_2)^2, & 0 < z \leqslant \tfrac{1}{2} z_2 \\ -m\omega_{z_1}^2 (z + \tfrac{1}{2} z_1)^2, & -\tfrac{1}{2} z_1 < z \leqslant 0 \\ 0, & \text{elsewhere.} \end{cases} \tag{7.39}$$

TABLE 7.1
Strengths of non-central components in the Nilsson potential

	Rare-earths	Actinides
κ_p	0·0688	0·0577
κ_n	0·0638	0·0635
μ_p	0·558	0·65
μ_n	0·491	0·325

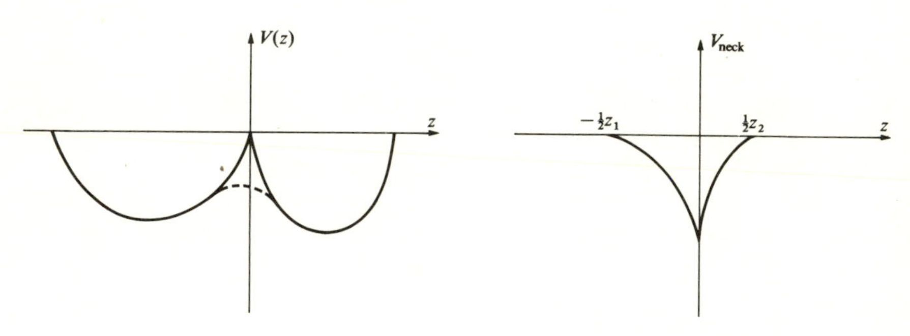

FIG. 7.8. The neck potential

In first-order perturbation theory this will yield a shift in the single-particle levels

$$\Delta E_{\text{neck}} = \langle \Psi \mid V_{\text{neck}} \mid \Psi' \rangle = \int_{\frac{1}{2}z_1}^{\frac{1}{2}z_2} Z(z)\, V_{\text{neck}}\, Z'(z)\mathrm{d}z. \tag{7.40}$$

Calculations yield this correction to be less than 1 per cent of the unperturbed eigenvalue $n_z \hbar\omega_z$ for $z_0 < 5$ fm rising to a few per cent at fission.

Using the single-particle states defined above we now compute the shell corrections and pairing energies as discussed in Chapter 6 and couple them to the liquid-drop potential energies illustrated in Fig. 7.5. The results are represented in the contour plot of Fig. 7.9. Following the fission path two new features are immediately obvious.

(1) There is now a double-humped barrier on the road to fission (Clark 1971). The fissioning process begins symmetrically across a barrier 5–6 MeV high until it reaches a second minimum.

(2) From the second minimum the fission path becomes asymmetric through a barrier 6–7 MeV high until at scission the asymmetry in the mass fragments is given by $p = A_2/A_1 \simeq 1{\cdot}4$, in excellent agreement with the observed yield illustrated in Fig. 2.14 (p. 19).

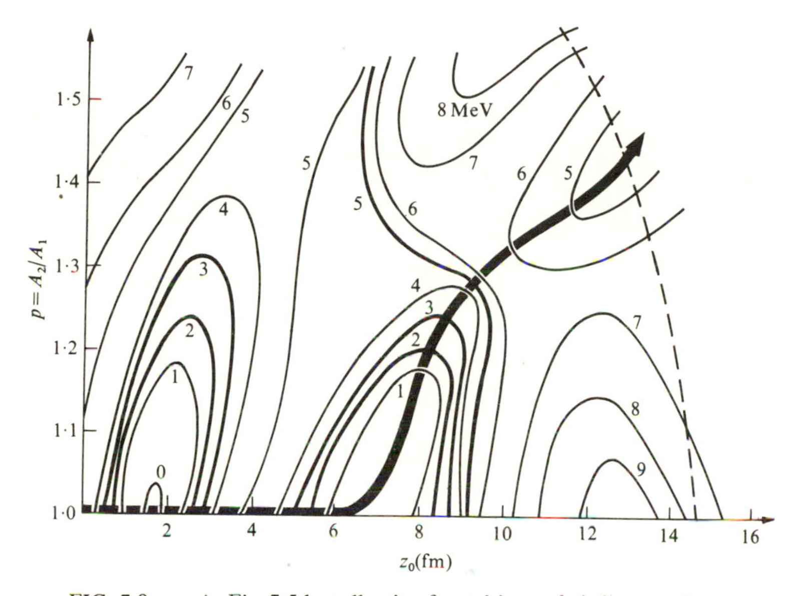

FIG. 7.9. As Fig. 7.5 but allowing for pairing and shell corrections.

In Fig. 7.10 the potential energy contours calculated for ^{256}Fm are plotted. This shows that for p from 1 to ~ 1·4 the height of the fission barrier is almost constant at 4–5 MeV and thus corresponds to a broad single-humped yield centred on symmetric fission. There is now no second minimum. Fig. 7.11 shows the energy contours for ^{226}Ra. Here we see that the symmetric path and asymmetric path barriers differ in height by only 2 MeV and thus are favourable to the production of the three-humped mass yield illustrated in Fig. 2.14. Finally, in ^{208}Pb we obtain the potential energy contours reproduced in Fig. 7.12, which favours symmetric fission. Being a very stable doubly magic nucleus the fission barrier in ^{208}Pb is correspondingly high at ~ 20 MeV. This should be compared with the energy of the giant dipole resonance at 10–20 MeV (see the comments on photo-fission in Chapter 2).

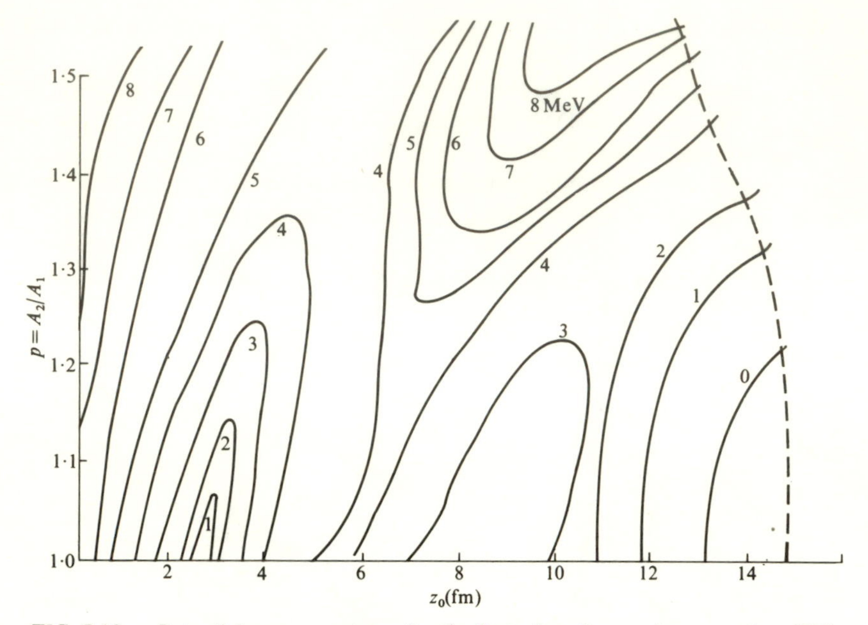

FIG. 7.10. Potential-energy contours for the fissioning of a very heavy nucleus ^{256}Fm.

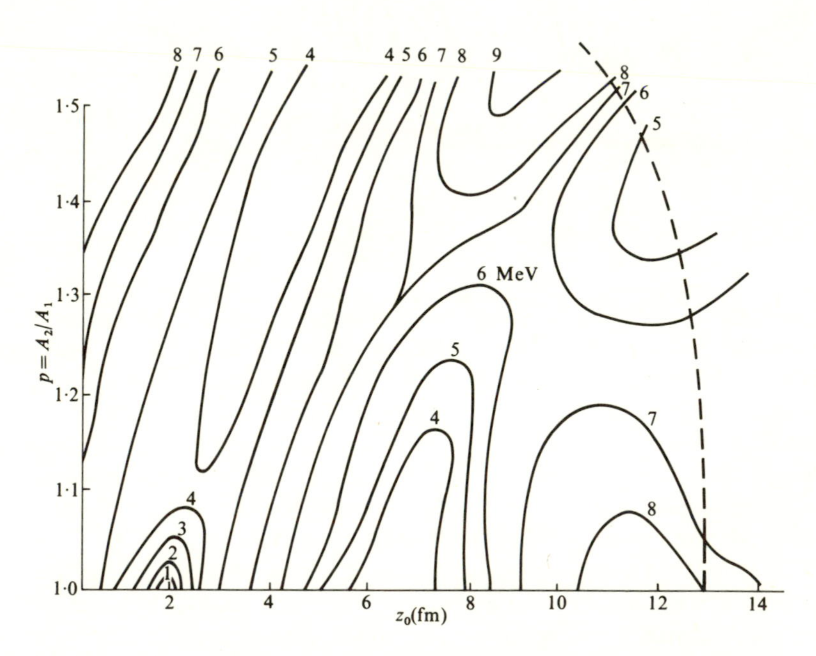

FIG. 7.11. Potential-energy contours for the fissioning of a light actinide ^{226}Ra.

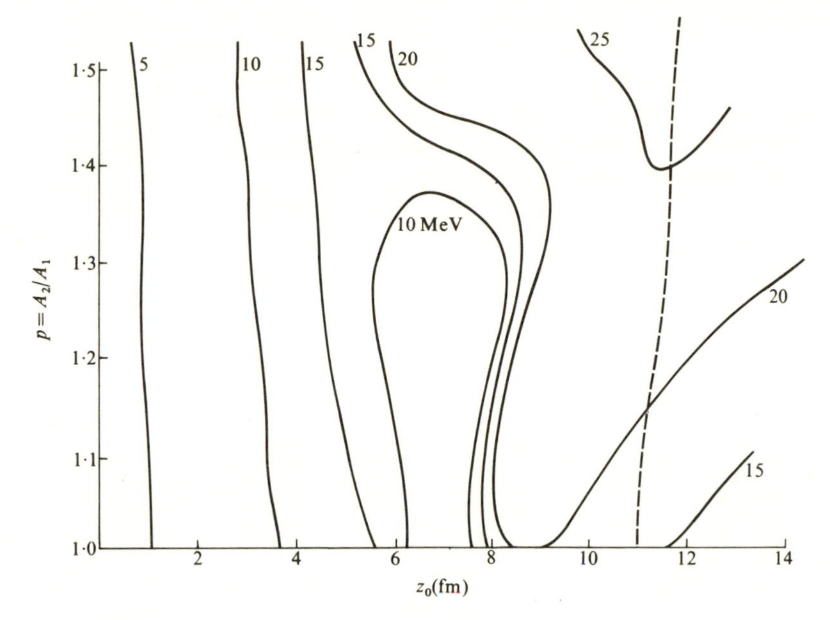

FIG. 7.12. Potential energy contours for the fissioning of ^{208}Pb.

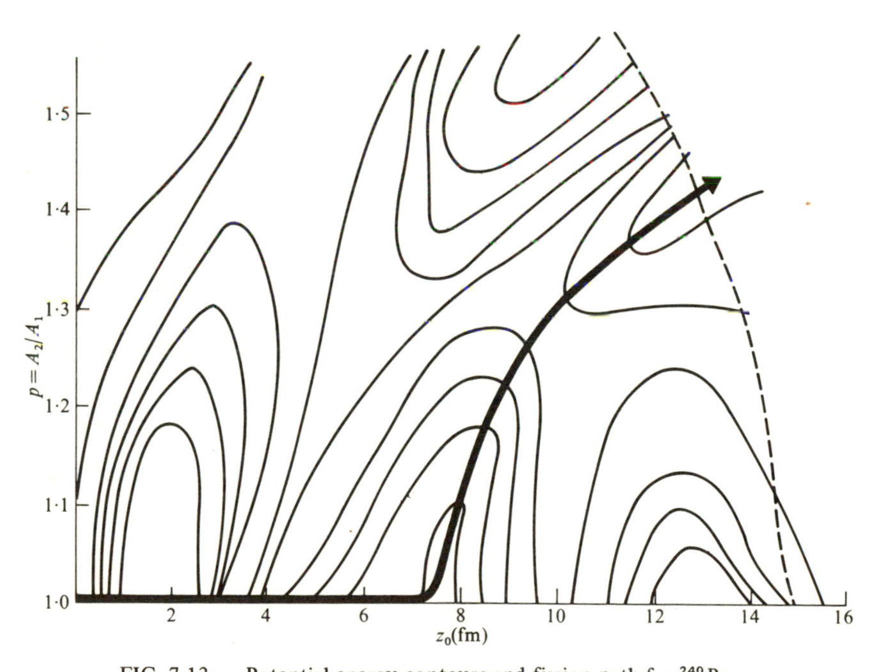

FIG. 7.13. Potential-energy contours and fission path for ^{240}Pu.

The results described above are extremely encouraging and suggest that we can adequately describe the main systematics of the fission of heavy nuclei. There remain at least three points that require further study: (1) the role of the second minimum observed in some fission barriers; (2) the detailed calculation of actual fission yields and fission lifetimes. (3) the fission properties of odd-mass isotopes.

To consider the first point let us study the case of ^{240}Pu. The potential-energy contours are reproduced in Fig. 7.13, and Fig. 7.14 shows the cross-section through the potential energy surface cut by the fission path. If this

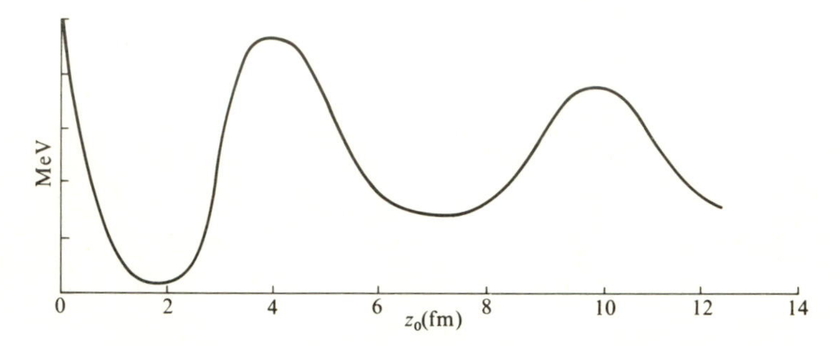

FIG. 7.14. Fission barrier corresponding to Fig. 7.13.

minimum is deep enough we can expect the nucleus to become trapped for a period in this secondary potential well. This metastable shape isomer is inhibited by the potential well from gamma-decaying back to the ground state or fissioning into two fragments and is seen therefore as a long-lived excitation of ^{240}Pu at 2–3 MeV. Since it represents a highly deformed configuration of ^{240}Pu it will have a large moment of inertia, and hence may support an extremely compressed rotational band. The members of this band lying within the potential well will have the choice of gamma-decaying down to the isomeric ground state or spontaneously fissioning. Thus the existence of an isomeric state will carry two signatures. There will be an anomalously long spontaneous-fission half-life and anomalously enhanced E2 transition to it from the excited rotational states built on it. Recently such states have been found in ^{240}Pu (see Fig. 7.15). We thus have two rotational bands in ^{240}Pu, one built on the deformed ground state and the other on the more highly deformed fission isomer. If we identify these with the two minima in Fig. 7.14 we can calculate the effective moments of inertia, as discussed in Chapter 5, using the single-particle states calculated as described above. In Fig. 7.16 we compare these calculated moments of inertia with those deduced from the observed excitation energies in the two rotational bands in Fig. 7.15. The small discrepancy between the calculated moments of inertia and those observed which is revealed in Fig. 7.16 is within the uncertainty in the strength of the pairing interaction.

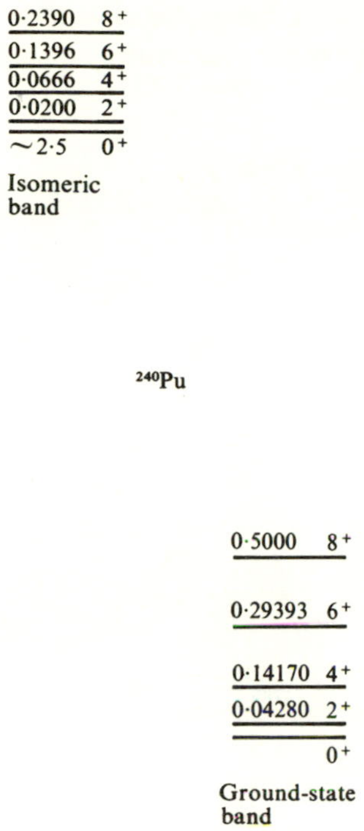

0·2390 8+
0·1396 6+
0·0666 4+
0·0200 2+
~2·5 0+
Isomeric band
240Pu
0·5000 8+
0·29393 6+
0·14170 4+
0·04280 2+
0+
Ground-state band

FIG. 7.15. Observed ground-state and isomeric-state rotational bands in ^{240}Pu.

300
250
200
150
100
50
2
4
6
8
10
12
14
z0(fm)
Ir
Ic
Ic+p
A
B

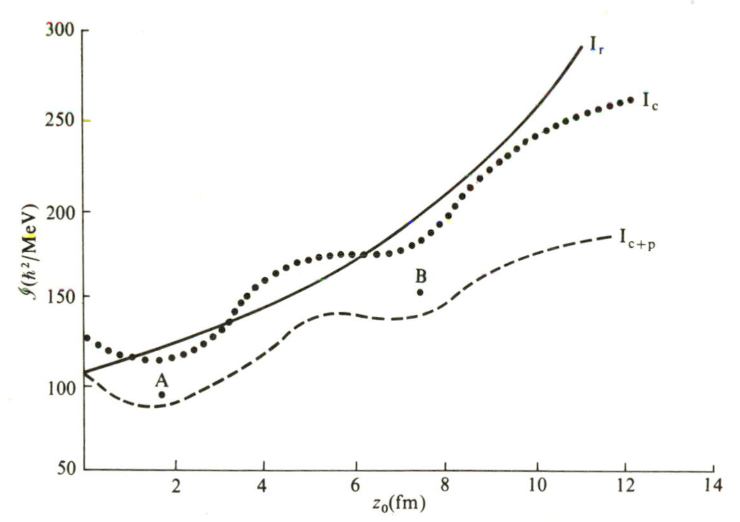

FIG. 7.16. Calculated moments of inertia as a function of centre separation in ^{240}Pu. I_r, the irrotational liquid drop; I_c, the cranking model; I_{c+p}, the cranking model plus pairing. A and B correspond to the ground-state and isomeric-state moments of inertia deduced from Fig. 7.15.

In order to calculate specific fission yields we shall assume that these are proportional to the fission barrier-penetration probabilities as given by the WKB approximation (see Fig. 7.17)

$$p = \exp\left[-2\int_{z_1}^{z_2}\sqrt{\left\{\frac{2B_Z(z)\,(V(z)-E)}{\hbar^2}\right\}}\mathrm{d}z\right]. \tag{7.41}$$

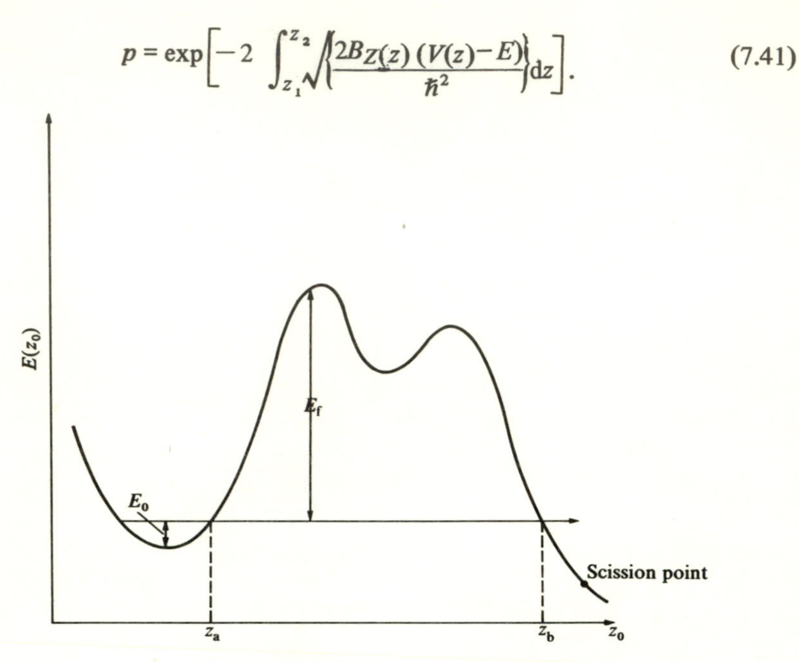

FIG. 7.17. Fission-barrier penetration

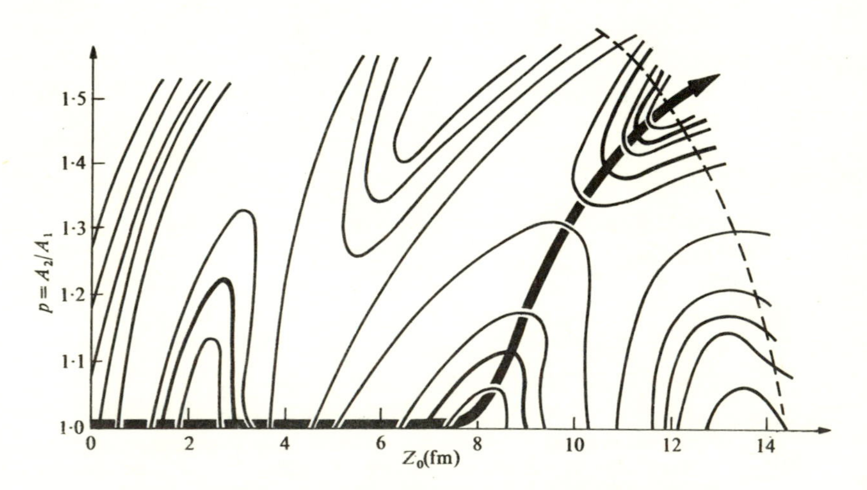

FIG. 7.18. Potential-energy contours and fission path for ^{242}Cm.

Fig. 7.18 shows the potential-energy contour plots for ^{242}Cm, and Fig. 7.19 compares the computed fission yields with those observed. In eqn (7.41) the only undefined quantity is the effective mass B_Z. This is given by

$$B_Z = 2 \sum_i \frac{|\langle 0 | \frac{\partial}{\partial z} | i \rangle|^2}{\epsilon_i - \epsilon_0}, \tag{7.42}$$

by the same argument that gives us the cranking formula (4.33). A plot of B_Z is shown in Fig. 7.20, and we note that as we would expect it tends asymptotically to the reduced mass of the two fragments $\mu = M_1 M_2/(M_1 + M_2)$, here calculated for the most probably yield.

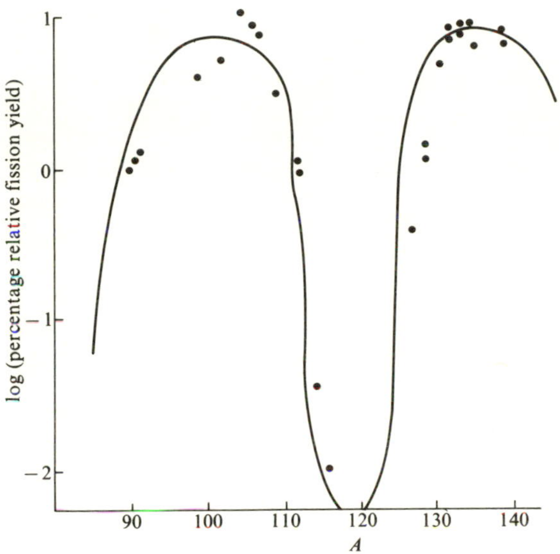

FIG. 7.19. A comparison of observed relative fission yields for ^{242}Cm compared with those calculated in the WKB approximation from Fig. 7.18.

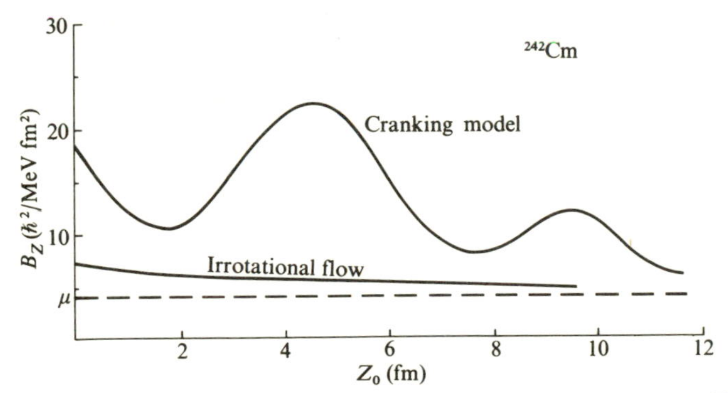

FIG. 7.20. Calculated inertia parameters for different centre separations in ^{242}Cm.

Finally, we turn to the difference in spontaneous fission behaviour observed in even- and odd-mass isotopes. Fig. 7.21 compares typical potential barriers calculated for neighbouring even- and odd-mass isotopes. In the ground states

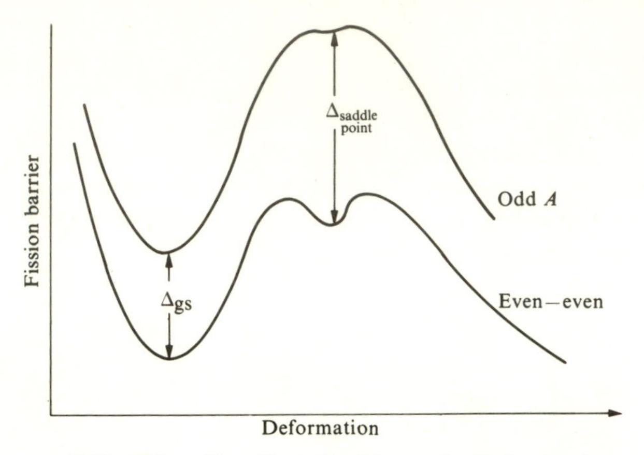

FIG. 7.21. The effect of pairing on the fission barrier.

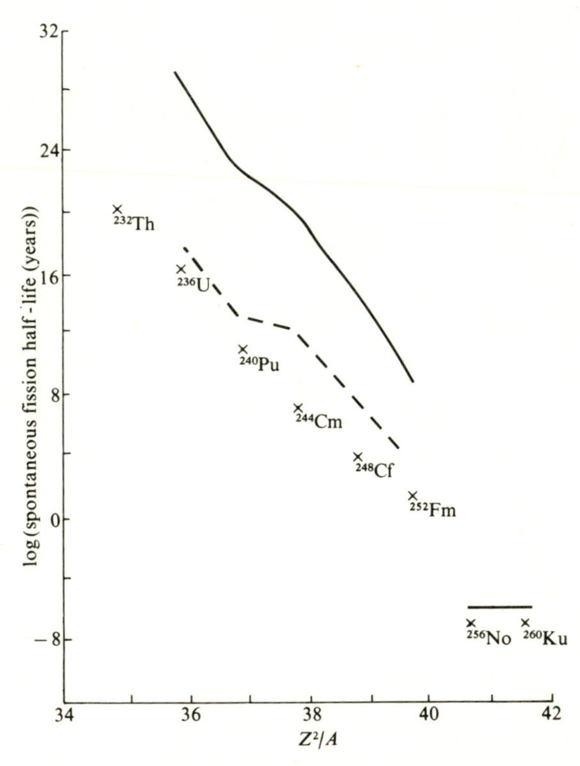

FIG. 7.22. Observed spontaneous fission half-lives compared with those calculated.

the pairing energy produces a gap Δ. The strength of the pairing correlations increase with deformation, and the presence of the extra particle in the odd-mass isotope destroys the shell effect leading to the secondary minimum. The result of these two effects is a much enhanced fission barrier in the odd-mass isotope, leading to a corresponding hindrance of the spontaneous fission process. We conclude our discussion of fission with Fig. 7.22, in which we compare calculated spontaneous fission half-lives with those observed.

Turning to alpha-decay, as stated at the beginning of the chapter, this is simply a special case of fission. However, the alpha particle is so compact and light compared with the parent nucleus that obvious approximations which greatly simplify the calculations suggest themselves. Since most of the heavy even–even nuclei spontaneously emit alpha particles in preference to any other form of

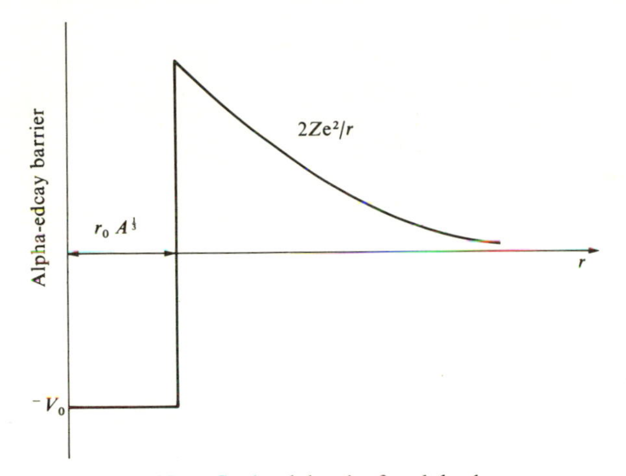

FIG. 7.23. Coulomb barrier for alpha decay.

decay we may assume that inertial effects are small and replace B_Z in eqn (7.41) by the reduced mass, which is approximately equal to the mass of the alpha particle. Secondly, since the size of the alpha particle is comparable with the surface thickness of the parent we may replace the complicated fission barrier by the approximate potential illustrated in Fig. 7.23.

Further reviews of material discussed in this chapter are to be found in Fong (1964), Irvine (1972), Rutherford, Chadwick, and Ellis (1930), Wheeler (1955), and Wilets (1964).

8

SUPERHEAVY NUCLEI

Until now we have concentrated on explaining observed phenomena in the region of heavy nuclei. We now enter the region of speculation. The first problem we shall discuss concerns the possible existence of nuclei much heavier than those already observed (Brack *et al* 1972; Johansson, Nilsson, and Szymanski 1970; Scharff Goldhaber 1957; Wheeler 1955).

Fig. 2.3 (p. 8) indicates that throughout the heavy-mass region the binding energy per particle decreases steadily with increasing mass number. The heaviest elements so far artificially manufactures, i.e. $A = 105, 106$, have half-lives of the order 1 s or less. The only mechanism that might halt this decline in binding energy is the shell correction effect. We refer to Fig. 6.4 (p. 117), where we notice that the low density of states in the spherical configuration corresponds to the observed magic numbers, and we note that after ^{208}Pb the next doubly magic nucleus is predicted to be $\sim {}^{298}114$. Fig. 8.1 shows the single-particle levels for particles moving in the potential well

$$V(r) = \tfrac{1}{2}m\omega^2 r^2 - \kappa\,\hbar\omega(2\,\mathbf{l}\,.\,\mathbf{s} + \mu\,\mathbf{l}^2). \tag{8.1}$$

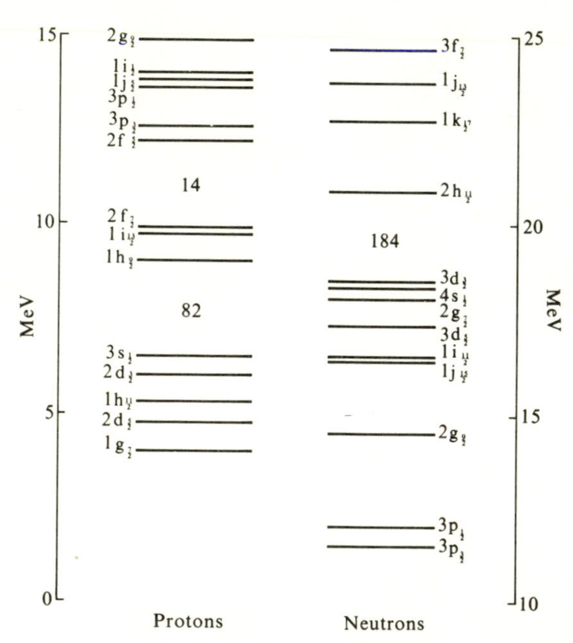

FIG. 8.1 Single-particle energies in the superheavy region.

The parameters κ and μ are chosen to reproduce the single-particle spacings in the rare-earth nuclei and the actinides (see Table 7.1) and are then linearly extrapolated to the heavier-mass region. We see that the proton magic numbers occur at $Z = 50$, 82, 114, and 164, while the neutron numbers occur at $N = 82$, 126, 184, and 196, possibly also 230 and 292. However, in order to observe a stable doubly magic nucleus we require more than simply the correct number of neutrons and protons. We require that the Fermi energies of the neutrons and protons should be closer together than the single-particle spacings of either the neutron or proton levels, otherwise beta decay will occur. If we assume that the charge is uniformly distributed throughout a sphere of radius $r_0 A^{\frac{1}{3}}$ then the Coulomb potential experienced by protons will be

$$\begin{aligned} V_C &= \frac{Ze^2}{r}, \quad r \geqslant R \\ &= -\frac{Ze^2}{R^3}(r^2 - 2R^2), \quad r < R, \end{aligned} \tag{8.2}$$

and in order to obtain a true comparison of the neutron and proton levels this should be added to the nuclear potential (8.1) in computing the proton single-particle levels. For the present we shall simply estimate the effect on a proton at the nuclear surface; its energy will be shifted upwards by $\sim Ze^2/r_0 A^{\frac{1}{3}}$. If we insert this Coulomb shift into Fig. 8.1 and take for $h\omega$ the conventional form $\sim 41\, A^{-\frac{1}{3}}$ MeV, then we find the most plausible candidates for doubly magic superheavy nuclei are the elements $^{298}114$, $^{310}114$, and $^{456}164$.

Given the single-particle states of (8.1) and (8.2) we can now calculate the shell corrections and pairing energies exactly as before. Fig. 8.2 shows a continuation of Fig. 6.5 (p. 121) extrapolated into the superheavy region. This figure suggests that element $^{298}114$ is most likely to be the next doubly magic nucleus

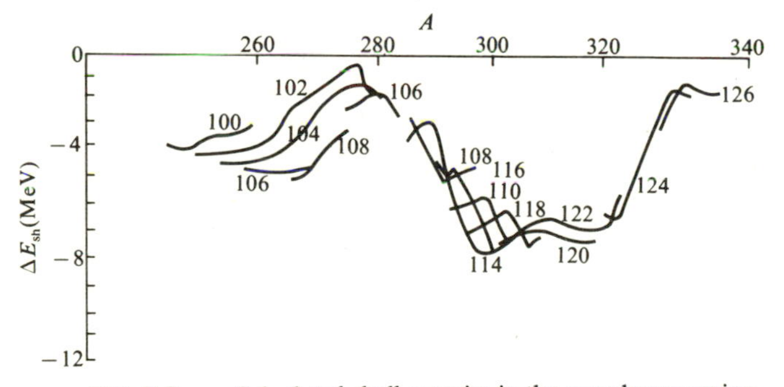

FIG. 8.2. Calculated shell energies in the superheavy region.

after ^{208}Pb. Having established the possible existence of a doubly magic configuration there next arises the question of its stability. For this we require to know the mass of the superheavy nucleus relative to the sum of the masses of decay products and the structure of the decay barriers. In Fig. 8.3 we illustrate the limits of nuclear stability predicted by the semi-empirical mass formula, including a pairing term but not allowing for shell corrections.

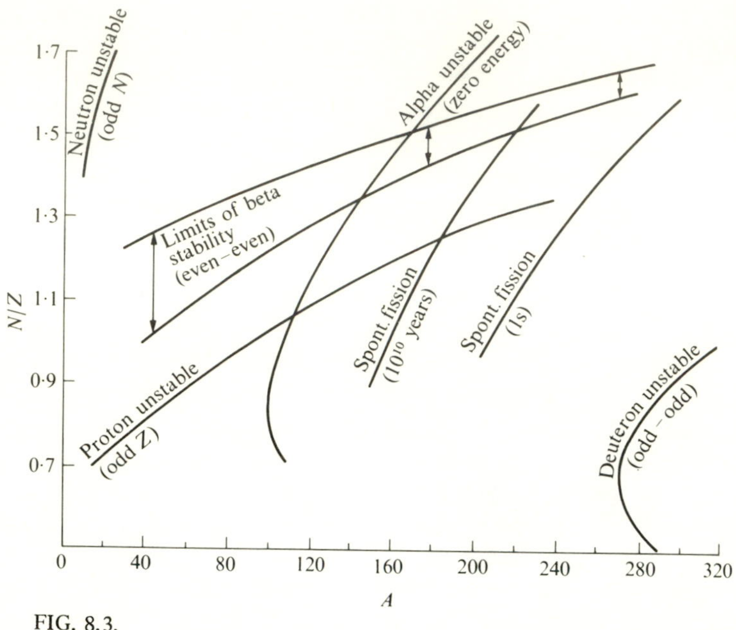

FIG. 8.3.

We should expect that the pairing effects will rule out the possibility of beta decay from an even–even doubly closed-shell nucleus. However, even–even nuclei in the superheavy region may have extremely short spontaneous fission half-lives. The spontaneous fission in inhibited in the odd-mass nuclei which may have short beta-decay half-lives. Thus which factor is going to limit the stability of the super-heavy nuclei will depend on detailed calculations.

Turning first to alpha decay the Geiger–Nuttall law (Fig. 2.8, (p. 15)) gives us an adequate relationship between the alpha-decay half-lives and the transition energies. In Fig. 8.4 we illustrated the calculated alpha-decay half-lives when the transition energy is taken from the extrapolated semi-empirical mass formula, and the result of correcting the energy for shell effects as represented in Fig. 8.2. As we would expect the shell effects tend to inhibit the alpha decay from the predicted closed-shell configuration. In the region around the closed shell $112 \leqslant Z \leqslant 116$ we see that the alpha-decay half-life is falling rapidly with increasing proton number. Further uncertainty is introduced by the fact that relatively small changes in the form of the single-particle potential can lead to a displacement of the calculated alpha-decay curve, although its shape is relatively insensitive to the potential. Since the curve is falling so rapidly these displacements can lead to large changes in the predicted alpha-decay half-life. Quoted values of the alpha-decay half-life of $^{298}114$ range from 10^{-5} years to 10^{5} years. Most calculations suggest that $^{296}112$ will have an alpha-decay half-life in excess of 10^{5} years.

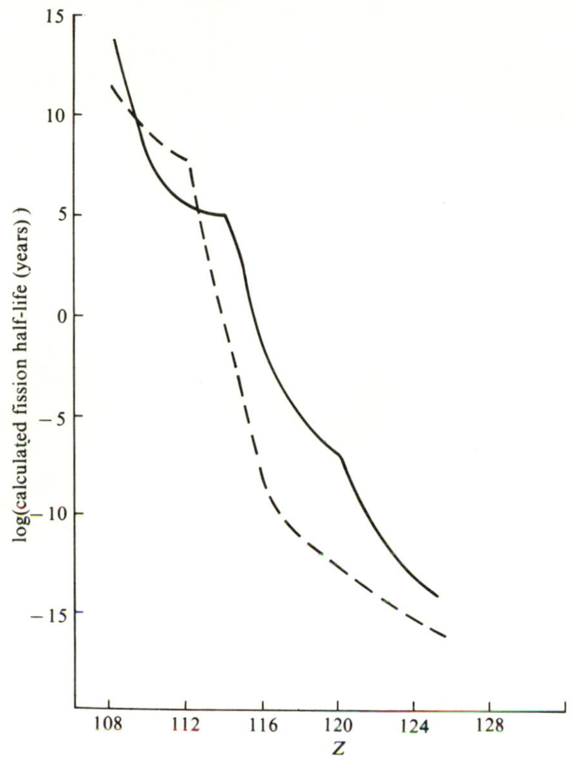

FIG. 8.4. Calculated alpha-decay half-lives in the superheavy region. The solid line is a liquid-drop model calculation and the broken line includes shell and pairing corrections.

Turning to spontaneous fission, Fig. 8.5. shows the fission barrier calculated in the two-centre model of Chapter 7 for our superheavy candidate $^{298}114$. We

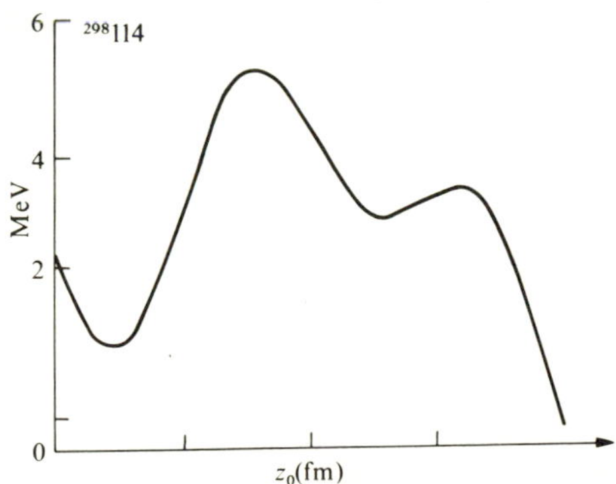

FIG. 8.5. The calculated fission barrier for superheavy element $^{298}114$.

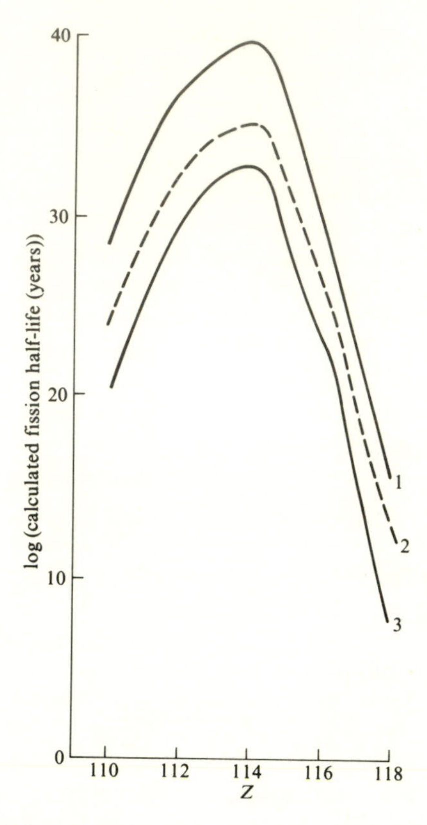

FIG. 8.6. Calculated fission half-lives in the superheavy region. The zero-point energies are assumed to be 1 MeV, 1·5 MeV, and 2 MeV for curves 1, 2, and 3 respectively.

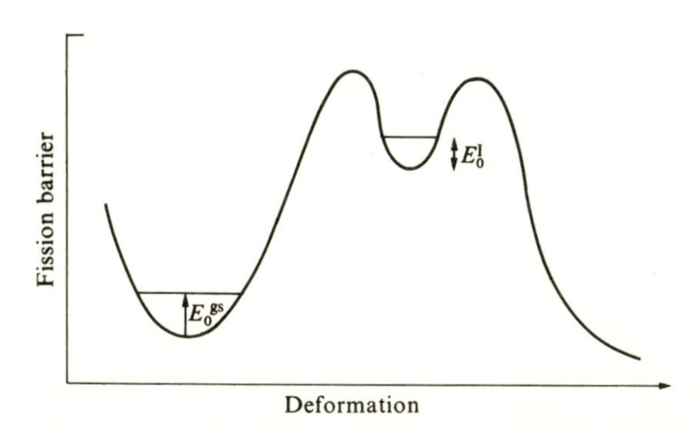

FIG. 8.7. Zero-point energies in the fission barriers.

see that the curve does not predict the existence of a stable shape isomer. This is due to the fact that there is no isomeric 'shell' for $Z = 114$. Fig. 8.6 shows the calculated fission half-lives of superheavy nuclei obtained in the barrier-penetration model of Chapter 7. We see that the spontaneous fission half-lives fall off extremely rapidly as we depart from the closed-shell $Z = 114$ configuration. There are two principal sources of uncertainty in the calculated fission half-lives; these are the effect of zero-point motion (see Fig. 8.7) and the pairing correlations. In Fig. 8.6 we have assumed that the ground-state well is approximated by a parabola and have taken the usual oscillator zero-point motion $\frac{1}{2}\hbar\omega$. This gives a zero-point energy of ~ 1 MeV in the nuclei represented in Fig. 8.6. Higher zero-point motion leads to shorter fission lives. We have also assumed bulk pairing as described in Chapter 4 with the strength chosen to reproduce the even–odd mass difference $\sim 12A^{-\frac{1}{2}}$ MeV observed in lighter nuclei. The assumption either of surface pairing or greater strength of pairing will serve to reduce the spontaneous-fission half-life. However, all our calculations suggest a spontaneous-fission half-life for $^{298}114$ in excess of 10^{20} years. Turning to Fig. 7.22 (p. 140) we see that our model consistently overestimates the spontaneous-fission lifetimes, but not in a manner which systematically increases with mass. Thus we may have some confidence that it is unlikely that the spontaneous-fission half-life of the ground state of $^{298}114$ will be so short that we shall not observe it.

We summarize our findings on the stability of superheavy nuclei in Fig. 8.8, from which we deduce that there is likely to be an island of stability for nuclei near $^{298}114$.

Several interesting search programmes have been undertaken in an attempt to observe superheavy nuclei. The Russians (Flerov 1972) have studied medieval stained glass windows, arguing that if superheavy elements exist naturally and are chemically similar to lead then lead glass should contain superheavy nuclei. If these nuclei then decay in the glass the glass will act like a nuclear emulsion and characteristic tracks should be observed. A group working in Britain (Marinov, Batty, Kilvington, Newton, Robinson, and Hemingway 1971) studied the tungsten beam stops from the 30 GeV CERN proton synchrotron. They estimated that a tungsten nucleus recoiling after stopping a 30 GeV proton would have sufficient energy to overcome the Coulomb barrier of a neighbouring tungsten nucleus and hence open the door to a possible fusion reaction; it was hoped that a stable superheavy element might be one of the decay products of such a reaction. More conventionally, terrestial ores have been studied (Flerov 1972) in great detail and nuclear emulsion studies of cosmic rays have searched for extra-terrestial superheavy elements. To date, no experimental evidence has been found for the existence of superheavy elements in nature. Hope is now pinned on producing them artificially in heavy-ion reactions using the next generation of heavy-ion accelerators. Unfortunately the favoured reactions are of the form

$$^{48}\text{Ca} + {}^{250}\text{Pu} \rightarrow {}^{298}114, \tag{8.3}$$

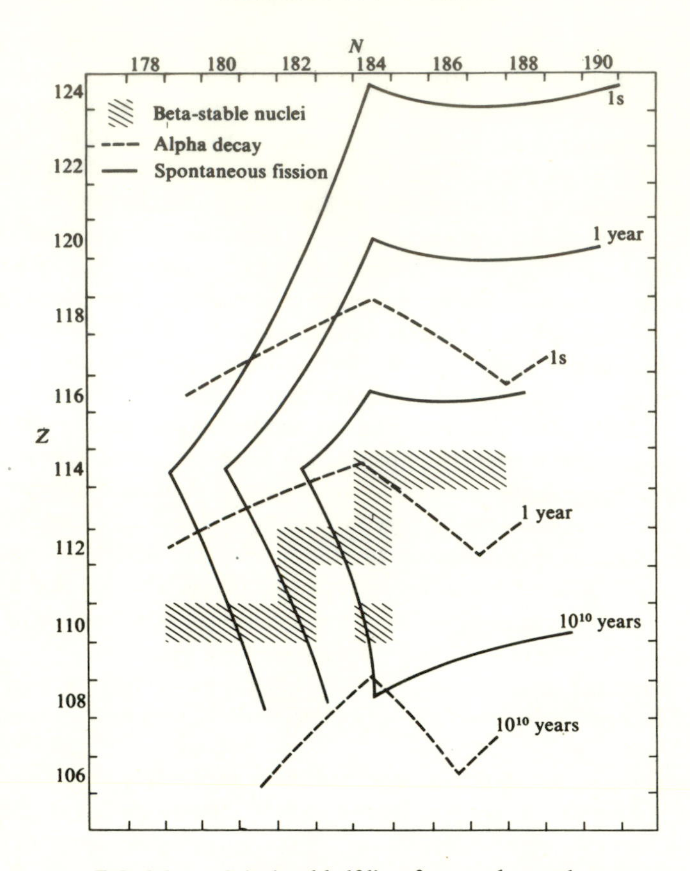

FIG. 8.8. Calculated half-lives for superheavy elements.

and neither the target nor the beam are practical possibilities. Possible reactions involve forming a compound state of two extremely heavy ions and searching for superheavy elements in the decay of the compound state. The limiting factor is then likely to be the superheavy element formation cross-section.

Chemically, superheavy elements should follow the regularities in properties exhibited in the periodic table (Fig. 8.9). Interesting theoretical possibilities arise when considering atoms with $Z > 137$, in which case the binding energy of the 1s electron exceeds its rest mass and spontaneous positron production becomes a possibility.

Shells																																																				
1s	H 1 $(1s)^1$																																																			He 2 $(1s)^2$
2s 2p	Li 3 $(2s)^1$	Be 4																																											B 5	C 6	N 7	O 8	F 9	Ne 10 $(2s)^2$ $(2p)^6$
3s 3p	Na 11 $(3s)^1$	Mg 12																																											Al 13	Si 14	P 15	S 16	Cl 17	Ar 18 $(3s)^2$ $(3p)^6$
3d 4p 4s	K 19 $(4s)^1$	Ca 20	Sc 21																																		Ti 22	V 23	Cr 24	Mn 25	Fe* 26	Co 27	Ni 28	Cu 29	Zn 30	Ga 31	Ge 32	As 33	Se 34	Br 35	Kr 36 $(3d)^{10}$ $(4p)^6$
4d 5p 5s	Rb 37 $(5s)^1$	Sr 38	Y 39																																		Zr 40	Nb 41	Mo 42	Tc 43	Ru 44	Rh 45	Pd 46	Ag 47	Cd 48	In 49	Sn 50	Sb 51	Te 52	I 53	Xe 54 $(4d)^{10}$ $(5p)^6$
5d 6p 6s 4f	Cs 55 $(6s)^1$	Ba 56	La 57																				Ce 58	Pr 59	Nd 60	Pm 61	Sm 62	Eu 63	Gd 64	Tb 65	Dy 66	Ho 67	Er 68	Tm 69	Yb 70	Lu 71	Hf 72	Ta 73	W 74	Re 75	Os 76	Ir 77	Pt 78	Au 79	Hg 80	Tl 81	Pb 82	Bi 83	Po 84	At 85	Rn 86 $(5d)^{10}$ $(4f)^{14}$
6d 7p 7s 5f	Fr 87 $(7s)^1$	Ra 88	Ac 89																				Th 90	Pa 91	U 92	Np -93	Pu 94	Am 95	Cm 96	Bk 97	Cf 98	Es 99	Fm 100	Md 101	No 102	Lr 103	104 $(6d)^4$	105 $(6d)^3$	106 $(6d)^4$	107 $(6d)^5$	108 $(6d)^6$	109 $(6d)^7$	110 $(6d)^8$	111 $(6d)^9$	112 $(6d)^{10}$	113 $(7p)^1$	114 $(7p)^2$	115 $(7p)^3$	116 $(7p)^4$	117 $(7p)^5$	118 $(6d)^{10}$
8p 6f 5g 8s 7d	119 $(8s)^1$	120 $(8s)^2$	121	122	123	124	125	126	127	128	129	130	131	132	133	134	135	136	137	138	139	140	141	142	143	144	145	146	147	148	149	150	151	152	153	154	155	156	157	158	159	160	161	162	163	164						
			The super actinides: the 6f and 5g shells play the role of the 5f shell in actinides																																																	
$9p_{1/2}$ $8p_{3/2}$ 9s	165 $(9s)^1$	166 $(9s)^2$																																											167 $(9p_{1/2})^1$	168 $(9p_{1/2})^2$	169 $(8p_{3/2})^1$	170 $(8p_{3/2})^2$	171 $(8p_{3/2})^3$	172 $(8p_{3/2})^4$

FIG. 8.9. The extended periodic table.

9

NEUTRON STARS

A main sequence star like the sun produces its energy by the fusion of light nuclei. This process is exothermic as long as the product nuclei are lighter than ^{56}Fe (see Fig. 2.3, p. 8). These reactions lead to an outward directed radiation pressure which together with the hydrodynamic pressure, balances the inward gravitational self-attraction. The mass density of a star during this phase is 1–10g cm^{-3}, and mean temperatures are in the range 10^5–10^7 K. The material of the star is in the form of a completely ionized plasma. When the fusion fuel is exhausted the radiation pressure ceases and the star collapses under its self-gravitational attraction (Clayton 1968; Harrison *et al.* 1965).

The collapse is halted, at least temporarily, by the electron Fermi pressure. In this state the mass density is in the range 10^4–10^{10} g cm^{-3}, the temperature is $\sim 10^7$ K, and the material of the star is in the form of a lattice of nuclei, principally ^{56}Fe with some heavier elements like ^{60}Co in the centre and lighter elements like Mn and Cr in the surface. At this temperature the nuclei are still completely ionized, but at this density the electron Fermi temperature is much greater than the ambient temperature. Thus the electrons form an almost completely degenerate highly relativistic electron gas. The pressure of this gas opposes the further collapse of the star. The electron density in this phase is much greater than in any terrestial metal, and the electron capture rate by the nuclei is correspondingly much higher. This has three effects: (1) the electron density is reduced and hence the Fermi pressure falls and the star contracts; (2) the charge number on the nuclei is decreased and they become more and more neutron-rich; (3) the neutrinos produced in the electron-capture carry energy away from the star.

One of the possible next stages in the evolution of the star is the neutron-star phase. In this state the mass density is in the range 10^{10}–10^{15} g cm^{-3} and the temperature is $\sim 10^7$ K. Most of the material is in the form of neutrons with ~ 10 per cent in the form of electrons and protons. The range of densities spans the central density of terrestial nuclei $\sim 3 \times 10^{14}$ g cm^{-3}. At these densities the nucleons are almost completely degenerate and the nucleon Fermi pressure balances the gravitational self-attraction. In many respects the star is now simply a giant nucleus of the mass of the sun $\sim 10^{31}$ g and the radius $\sim 10^5$–10^6 cm.

Fig. 9.1 illustrates the likely structure of a neutron star. Of the central baryon-soup region little can be said except that the density is $\gtrsim 10^{15}$ g cm^{-3}, corresponding to a neutron Fermi energy $\gtrsim$ the pion rest mass, and hence it is energetically favourable to have a baryon phase mixture (Baym, Bethe, and Pethick 1971; Canuto 1973–4).

We shall refer to the region $3 \times 10^{14}-10^{15}$ g cm^{-3} as the neutron-matter region. Here the material is in the form of a non-relativistic highly-degenerate neutron fluid. The density is high enough for protons and electrons to be almost entirely excluded, except at the extreme surface of this region. One interesting

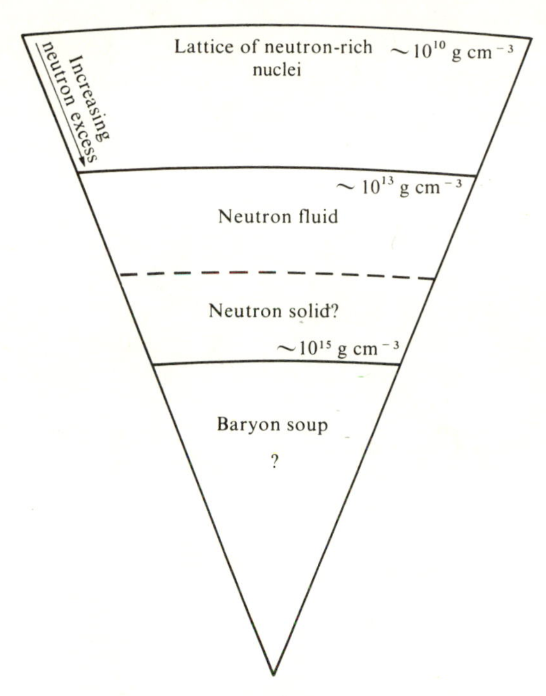

FIG. 9.1. Cross-section of neutron star.

question which arises in this region is, does the neutron fluid become a solid in the high-pressure region? Because of the similarity in form between the ^{3}He–^{3}He interatomic potential and the neutron–neutron nuclear interaction (see Fig. 9.2), it has been suggested that by a correspondence-of-states argument a neutron fluid under pressure should solidify.

The similarity in form between the two potentials exists between the $^1S_0$ partial wave interactions only. It is characterized by a strong short-range repulsion and a longer-range, much weaker attraction. In liquid ^{3}He the interaction is assumed to have the same form in all relative partial waves, while the neutron–neutron potential exhibits a strong exchange character, being in general much more repulsive in odd partial waves.

In liquid ^{3}He, under low pressure, typical effective $^1S_0$ interaction matrix elements are repulsive, and the binding energy is supplied by the ^3P state and higher partial wave interactions, which are all attractive. As the pressure increases, the polarization of the atoms giving rise to the van der Waal's forces increases; the

$^1S_0$ interaction becomes more repulsive, while the longer-range higher partial wave interactions become more attractive. It first becomes favourable to form dimers (virtual He_2 molecules) and then a solid. One interesting aspect of this is that the spins, which reside on the nuclei of the atoms, are virtually uncoupled in the solid, and hence the solid behaves like a classic paramagnet for which the spin entropy goes to zero with decreasing temperature more slowly than it does for the Fermi fluid. Thus the entropy of the phase is higher than for the liquid phase.

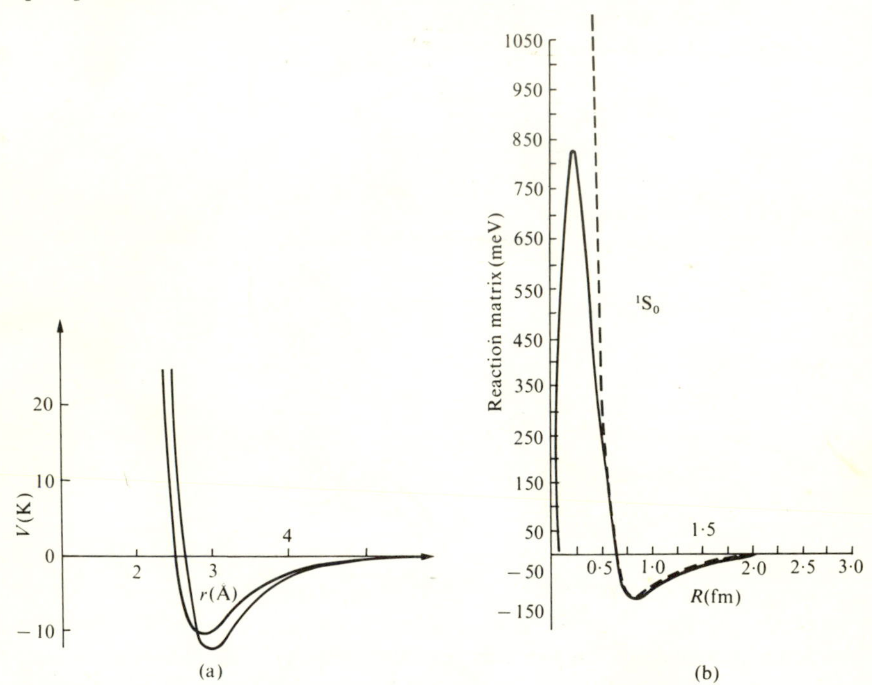

FIG. 9.2. A comparison between the ^3He interatomic potential and the neutron–neutron $^1S_0$ interaction. Two versions of the bare ^3He potential are given in (a). These should be compared with the dashed curve in (b) which represents the neutron–neutron $^1S_0$ interaction. The solid line is the effective Brueckner–Goldstone potential.

In neutron matter at normal nuclear densities the $^1S_0$ interaction is attractive while the ^3P interactions are zero or slightly repulsive. As the density is increased the $^1S_0$ interaction becomes less attractive and the ^3P interaction becomes rapidly more repulsive (principally because of the size of the $^3P_0$ core; the ^1D states become significant and are attractive, but they can never keep pace with the increasing repulsion from the P waves. Thus there is a significant difference between liquid helium, where saturation is dictated by the size of the repulsive core, and neutron matter, where saturation is provided principally by the exchange nature of the force. Therefore the dineutron, unlike the helium molecule, is unlikely to be formed by the central forces. There is in the neutron-matter case the tensor force in the $^3P_2-^3F_2$ channel which is of exactly the same form as the dipole–dipole interaction which give rise to van der Waal's forces. This may lead to the formation of solid neutron, although simple estimates probably favour a spin-ordered fluid. In the neutron case, unlike the situation in ^3He, the dipole moment and the nuclear spin are one and the same thing, and hence any neutron solid would be dominated by spin–spin correlations and would probably be spin-ordered. Thus the entropy of the solid phase would be expected to be lower than that in the liquid phase. It should be clear by now that simple corresponding-states arguments are going to tell us little about any solid phase of neutron matter.

In the neutron fluid the methods used to calculate the properties of nuclear matter might be thought to be valid. However, even here care must be taken since these calculations assume a 'low' density (see eqn (3.62), p. 44) and should not be trusted beyond 3×10^{14} g cm^{-3}. The techniques for treating dense Fermi fluids are known, but are computationally so complex that no serious application to neutron matter has so far been atempted.

We are left with the region of the nuclear lattice and the top of the neutron fluid, which we shall now discuss in some detail.

In this region we have a lattice of nuclei characterized by a mass number A and charge number Z immersed in a neutron gas with possibly an extremely small number of protons and a balancing number of electrons to ensure over-all charge neutrality. In the extreme low-density region the lattice nuclei will be predominately ^{56}Fe and the surrounding gas will consist almost entirely of electrons. As the density increases the electron gas density will decrease and the nuclei will become more and more neutron-rich. Some neutrons will escape to form the neutron gas, and eventually the nuclei will dissolve in the neutron gas, to form the beginning of the neutron-matter region. The independent variables are A, Z, the density of nuclei n_N, the neutron gas density n_n and the relative abundance of protons y. The electron density n_e is then fixed by the requirement of electrical neutrality. We require to calculate the energy density of such a system minimized with respect to these variables subject to the constraints of fixed baryon number and fixed mass density. This will then yield an energy versus mass density curve from which an equation of state may be deduced.

We may write the total energy density as a function of five terms

$$E_t(A, Z, n_N, y) = n_N E_N + \bar{n} E_g + n_e E_e + n_N E_L + \{n_n m_n + n_N(Zm_p + Nm_n) + n_e m_c + y n_n m_p\} c^2, \quad (9.1)$$

where E_N is the binding energy per nucleus, E_g is the nucleon gas energy per nucleon, the nucleon density is given by $\bar{n} = n_n(1 + y)$, E_e is the free electron gas energy per particle, and E_L is the usual lattice energy for a solid; the fifth term represents the total rest-mass energy.

The nuclear binding energy has been discussed at some length in Chapter 6. However, that analysis is deficient in the present case in two respects. First, the effective interactions between nucleons in the nuclei will be modified by the proximity of the ambient nucleon gas. This will lead to a modification of the volume term in the liquid-drop energy which will now tend to be smaller. Second, the presence of the nucleon gas will severely reduce the surface energy, since there is now no clear-cut surface to the nuclear material. Hence the dependence of E_N on n_n and y in eqn (9.1). The modification of the effective interaction between nucleons may be expected to be a relatively small effect, and here we shall neglect it. We propose to calculate the nuclear binding energy as described in Chapter 6, but to remove from the liquid-drop energy the phenomenological surface energy of section 3.2 and replace it by a surface energy calculated to take account of the surrounding nucleon gas.

We shall denote the neutron density inside the nucleus by ρ_n, the proton to neutron ratio by $x = Z/N$, and assume that the charge (i.e. proton) density is proportional to the neutron density, i.e. $\rho_p = \rho_n$. Thus the total nucleon density in the nucleus will be

$$\bar{\rho} = \rho_n(1 + x). \tag{9.2}$$

We define neutron and proton radii for the nuclei by

$$\begin{aligned} Z &= 4\pi \int_0^{R_p} x\,\rho_n(r)\,r^2\,dr, \\ N &= 4\pi \int_0^{R_n} \rho_n(r)\,r^2\,dr, \end{aligned} \tag{9.3}$$

where

$$\rho_n(R_n) = n_n \tag{9.4}$$

and

$$\rho_n(R_p) = (y/x)n_n. \tag{9.5}$$

Since the proton relative abundance in the nucleon gas is expected to be much smaller than the proton relative abundance in the nuclei we have $y \ll x$ and $R_p > R_n$. We have now defined the neutron and proton densities throughout the region.

According to Thomas–Fermi theory the surface energy of ordinary nuclei composed of symmetric nuclear matter, i.e. $N = Z$, is given by

$$E_s = \frac{B}{\rho_0} \int (\Delta\bar{\rho})^2\,d\mathbf{r}, \tag{9.6}$$

where ρ_0 is the central saturation density and B is constant $\simeq 24$ MeV fm^2 related to the strength and range of nuclear forces. In the unsymmetric situation that we have, $x = 1$, then (9.6) is generalized to

$$E_s = \frac{1}{\rho_0} \int \mathrm{d}\mathbf{r} \{B_n (\nabla\rho_n)^2 + B_{pp} (\nabla\rho_p)^2 + 2B_{np}\nabla\rho_p . \nabla\rho_p\}, \qquad (9.7)$$

Because of the charge symmetry of nuclear forces $B_{nn} = B_{pp}$ coming only from the isotopic triplet components of the nuclear force. Note there are no Coulomb effects since these are already included in E_{ld}. The neutron–proton strength B_{np} is $\sim 2B_{nn}$ since this includes isotopic singlet interactions including the dominant $^3S_1$ state force. Thus

$$E_s \simeq \frac{B_{nn}(1 + 4x + x^2)}{\rho_0} \int_{|r| \leqslant |R_n|} (\nabla\rho_n)^2 \, \mathrm{d}r + \frac{B_{nn}}{\rho_0} \int_{|R_n| \leqslant r \leqslant |R_p|} (\nabla\rho_p)^2 \, \mathrm{d}r, \qquad (9.8)$$

and we have defined E_N.

The nucleon gas energy E_g can be calculated by standard nuclear-matter techniques in the simplifying approximation $y \ll 1$.

The free-electron gas energy is given by the well-known expression

$$E_e = m_e c^2 g(s)/8s^2, \qquad (9.9)$$

where

$$s = \frac{\hbar}{m_e c} (3\pi^2 n_e)^{\frac{1}{3}} \qquad (9.10)$$

and

$$g(s) = 3s(2s^2 + 1)\sqrt{(s^2 + 1)} - 3\ln\{s + \sqrt{(s^2 + 1)}\} - 8s^2. \qquad (9.11)$$

In the Wigner–Seitz approximation the lattice energy is given by

$$E_L = -\frac{9}{10} \frac{Z^2 e^2}{R_l} \left(1 - \frac{1}{3} \frac{r_p^2}{R_l^2}\right), \qquad (9.12)$$

where R_l is the average lattice spacing between the nuclei and r_p^2 is the mean-square charge radius of the nucleus

$$R_l = \left(\frac{3}{4\pi n_N}\right)^{\frac{1}{3}}, \quad r_p^2 = \frac{5}{3Z} \int r^2 \rho_p \, \mathrm{d}\mathbf{r}. \qquad (9.13)$$

The total energy (9.1) is now minimized with respect to the independent variables A, Z, n_n, and y at a given over-all nucleon density n_0,

$$n_0 = A n_N + \bar{n}, \qquad (9.14)$$

and thus we obtain E_t as a function of density n_0. The optimum values of A, Z, n_n and y give us the composition of the neutron-star matter at that density.

Since we are effectively at zero absolute temperature the Helmholtz free energy is essentially the same as the internal free energy, and hence we have the pressure

$$P(n_0) = n_0 \frac{dE_t(n_0)}{dn_0} - E_t(n_0). \tag{9.15}$$

The radius of the star is at a point where the pressure in eqn (9.15) exactly balances the inward gravitational attraction. This is given by the total mass density.

$$n_{mass} = E_t/c^2. \tag{9.16}$$

A number of calculation have been carried out which are minor variations on the above theme. Typical results are contained in Figs. 9.3 and 9.4. In Fig. 9.3

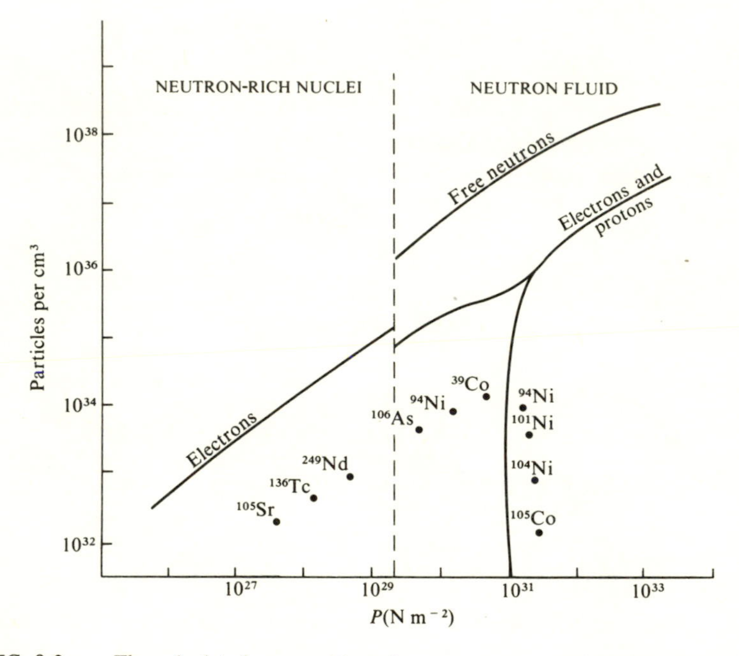

FIG. 9.3. The calculated composition of a neutron star as a function of pressure. There is a first-order phase transition at $P \sim 2 \times 10^{29}$ Nm^{-2}.

the composition of the neutron star as a function of pressure is indicated. At a pressure of $\sim 2 \times 10^{29}$ N m^{-2} there is a first-order phase transition from the lattice region to the neutron-fluid region. The equation of state for the neutron-star matter is shown in Fig. 9.4, and Fig. 9.5 shows how the neutron and proton densities change with increasing pressure.

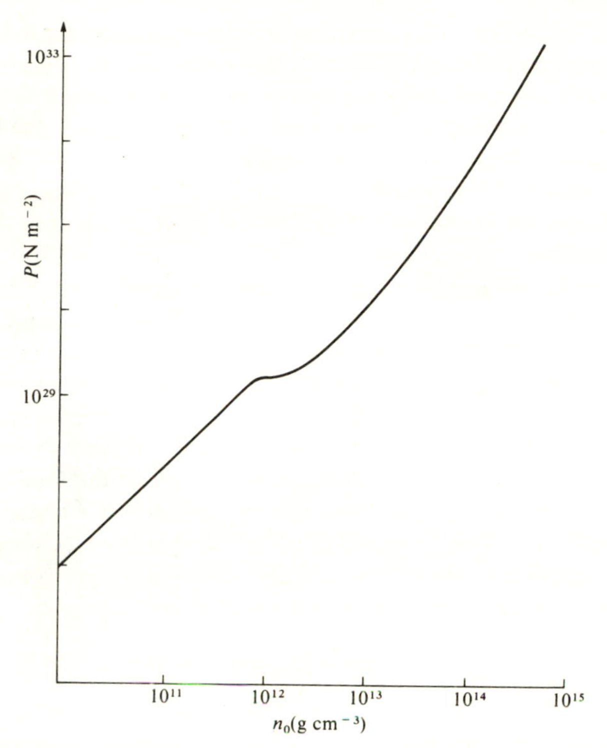

FIG. 9.4. The calculated equation of state for neutron-star matter. There is a discontinuity in the slope of the curve at $P \sim 2 \times 10^{29}$ $\mathrm{Nm^{-2}}$ corresponding to a first-order phase transition.

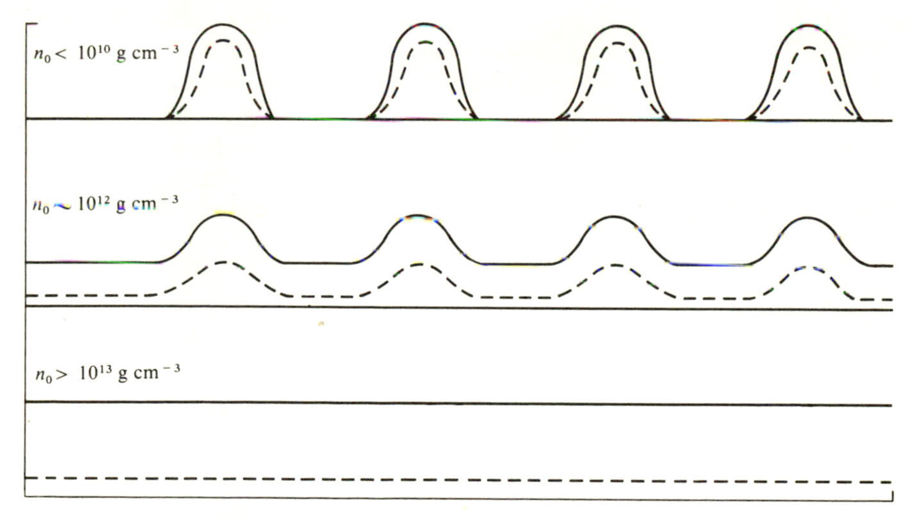

FIG. 9.5. Varying neutron distributions ———— and proton distributions . . . . as a function of density.

Clearly the above model for neutron-star matter could be improved in many details. Outstanding questions include: Is the neutron fluid a superfluid? If so, is it so through the complete range of density? Does the neutron fluid solidify under pressure, and if so is it a spin-ordered solid? Is the proton gas in the lattice region a superconductor? This is of great significance for the surrounding magnetic field of the neutron star, especially if the star has a ferromagnetic core. These questions all concern the role of the nuclear pairing correlations in the various regions of the star.

We have so far neglected the effects of temperature which at $\sim 10^7$ K is much smaller than the relevant Fermi temperatures which are in the range $10^{11}-10^{12}$ K. However, this does not mean that the matter is all in the form of a normal degenerate Fermi material, after all superconductivity occurs at temperatures well below the corresponding electron Fermi temperature. What is necessary for correlations of this type is simply that the density of states at the Fermi surface be high enough. Such correlations do not significantly effect the total binding energy, but they can drastically effect other properties of the system. One interesting possibility is that the nuclei in the lattice region are extremely neutron-rich and perhaps extremely deformed. They could thus support rotational bands, and if the moment of inertia is high enough such that

$$3\hbar^2/\mathscr{I}_{\text{eff}} < kT \tag{9.17}$$

then the thermal energy would be stored principally in rotations of the lattice nuclei. These would be inhibited if the surrounding nucleon fluid was normal, but would not be inhibited if it were a superfluid.

One other contribution to the pressure which we have neglected is provided by the neutrinos released in the electron-capture by the lattice nuclei. This increases with increasing density so that there is a net outward flux of neutrinos. At neutron-star densities the mean free path of neutrinos is only a few meters, and hence they may make a substantial contribution to the pressure.

Finally, there are effects due to general relativity which we have totally ingored but which it has been suggested may be more significant than hitherto believed at densities only slightly greater than that of nuclear matter.

Here we will stop, although we have by no means reached the end . . . .

REFERENCES

Baym, G., Bethe, H. A., and Pethick, C. J. (1971). *Nucl. Phys.* **A175**, 225.
Belyaev, S. T. (1959). *Math.-fys. Meddr* **31**, No. 11.
Bethe, H. A. (1971). *A. Rev. nucl. Sci.* **21**, 93.
Brack, M., Damgaard, J., Jensen, A. S., Pauli, H. C., Strutinsky, V. M., and Wong, C. Y. (1971). *Rev. mod. Phys.* **44**, 320.
Brown, G. E. (1967). *Unified theory of nuclear models and forces.* North-Holland, Amsterdam.
Canuto, V. (1973–74). *Equation of state at ultra high densities.* NORDITA Lectures, Copenhagen.
Clark, D. D. (1971). *Physics today* , 23.
Clayton, D. D. (1968). *Principles of stellar evolution and nucleonsynthesis.* McGraw-Hill, New York.
Cohen, S. and Swiatecki, W. J. (1962). *Ann. Phys.* **19**, 67.
Davidson, J. (1968). *Collective models of the nucleus.* Academic Press, London.
De Shalit, A. and Talmi, I. (1963). *Nuclear shell theory.* Academic Press, New York.
Flerov, G. N. (1972). In *Proceedings of the European Conference on Nuclear Physics* (Aix en Provence), Vols, I and II. Journal de Physique, Paris.
Fong, P. (1964). *Statistical theory of nuclear fission.* Gordon–Breach, New York.
Green, A. M. (1965). *Rep. Prog. Phys.* 28, 113.
Harrison, B. K., Thorne, K. S., Wakano, M., and Wheeler, J. (1965). *Gravitation theory and gravitational collapse.* The University of Chicago Press, Chicago.
Holzer, P., Mosel, U., and Greiner, W. (1967). *Nucl. Phys.* **A138**, 241.
——— , (1971). *Nucl. Phys.* **A164**, 257.
Hyde, E. K., Perlman, I., and Seaborg, G. T. (1971). *The nuclear properties of the heavy elements,* Vols. 1, II, and III. Dover, New York.
Irvine, J. M. (1972). *Nuclear structure theory.* Pergamon Press, Oxford.
Johansson, T., Nilsson, S. G., and Szymanski, Z. (1970). *Annls. Phys.* (5), 377.
Lane, A. M. (1964). *Nuclear theory.* North-Holland, Amsterdam.
MacDonald, N. (1970). *Adv. Phys.* **19**, 371.
Marinov, A., Batty, C. K., Kilvington, A. I., Newton, G. W. A., Robinson, V. J., and Hemingway, J. D. (1971). *Nature, Lond.* **229**, 464.
Nilsson, S. G. (1955). *Math.-fys. Meddr.* **29**, No. 16.
Nix, J. R. and Swiatecki, W. J. (1965). *Nucl. Phys.* **71**, 1.
Nuclear Data B (1966). **1**, No. 5.
(1969). **3**, No. 2.
(1970). **4**, No. 6.
(1971). **5**, No. 3.
(1971). **5**, No. 6.
(1971). **6**, No. 3.
(1971). **6**, No. 6.
Rutherford, E., Chadwick, J., and Ellis, J. (1930). *Radiations from radioactive substances.* Cambridge University Press, New York.
Scharff-Goldhaber, G. (1957). *Nucleonics.* **15**, 122.
Sorensen, R. A. (1973). *Rev. mod. Phys.* **45**, 353.

Strutinsky, V. M. (1968). *Nucl. Phys.* **A122**, 1.
Wheeler, J. A. (1955). Nuclear Fission and Nuclear Stability. In *Niels Bohr and the development of physics.* Pergamon Press, London.
Wilets, L. (1964). *Theories of nuclear fission.* Clarendon Press, Oxford.

AUTHOR INDEX

Batty, C. 147
Baym, G. 150
Belyaev, S. T. 1, 102
Bethe, H. A. 44, 79, 150
Bohr, N. 120
Born, M. 82
Brack, M. 112, 141
Brown, G. E. 23, 40, 78
Brueckner, K. 44
Canuto, V. 150
Chadwick, J. 15, 141
Clark, D. D. 133
Clayton, D. 150
Cohen, S. 38
Damgaard, J. 112, 141
Davidson, J. 23, 48, 102
De Shalit, A. 23, 45
Ellis, J. 15, 141
Flerov, G. N. 147, 149
Fong, P. 141
Geiger, H. 14, 15, 144
Goldstone, J. 44
Green, A. M. 82
Greiner, W. 130
Harrison, B. K. 1, 150
Hemingway, J. D. 147
Hewish, A. 1
Holzer, P. 130
Hyde, E. K. 22
Irvine, J. M. 123, 40, 78, 141
Jenson, A. S. 112, 141
Johansson, T. 141
Kilvington, A. I. 147
Landau, L. 1
Lane, A. M. 92
Marinov, A. 147
Mosel, U. 130
Newton, G. 147
Nilsson, S. G. 48, 49, 50, 51, 53, 122, 132, 141
Nix, J. R. 1, 38
Nuttall, J. 14, 15, 144
Oppenheimer, R. 82
Pauli, H. C. 112, 141
Pethick, C. 150
Perlman, I. 22
Robinson, V. J. 147
Rutherford, E. 15, 141
Scharff-Goldhaber, G. 141
Seaborg, G. 22
Seeger, P. 27
Sorensen, R. A. 1, 102
Strutinsky, V. M. 112, 141
Swiatecki, W. J. 1, 38
Szymanski, Z. 141
Talmi, I. 23, 45
Thorne, K. 1, 150
Wakano, M. 1, 150
Wheeler, J. 1, 38, 112, 141, 150
Wilets, L. 122, 141
Zwicky, F. 1

SUBJECT INDEX

Adiabetic approximation 81, 128
alpha-decay 14, 15, 16, 30, 141, 144
asymmetric fission 16, 126

Back bending 102, 106
band crossing 110
barrier, coulomb 22, 112, 141
    fission 39, 129, 133, 138, 145
beta-decay 13, 147
binding energy 3, 8, 121
Born–Oppenheimer approximation 81
breathing mode 29

Centrifugal force 107
    stretching 108
coriolis force 107
correlations, nuclear 78
    quadrupole 90
    pairing 92
    short-range 43
corresponding states 151
Coulomb barrier 22, 112, 141
    energy 25, 116, 124
    induced fission 20
cranking model 87, 139, 158

Decay of heavy nuclei 3
deformed shell-model 48

Electron gas in neutron stars 153, 155
energy, binding 3, 8, 121
    Coulomb 25, 116, 124
    density in neutron star 153
    pairing 26, 93, 98, 108, 121, 140
    quasiparticle 99
    rotational 11, 37, 111, 137
    single particle 41, 46, 53, 117, 131, 141
    surface 25, 116, 124, 154
    symmetry 26

Fission 122
    asymmetry 16, 126
    barrier 39, 129, 133, 138, 146
    Coulomb induced 20
    effects of pairing, 140
    isomers 136
    liquid-drop model of 38, 122
    neutron induced 18
    photo induced 20
    potential energy contours 128
    spontaneous 16, 17, 125
    superheavy element 145
    two-centre model of 127
    yields 19, 139

Gap, pairing 93, 95, 98, 140
giant dipole resonance 86

Hartree–Fock theory 78

Individual particle model 40
isomer, fission 136
    shape 110, 120

Lattice, energy 153, 155
    nuclear 151
liquid drop model 22, 116
    and fission 38, 122
liquid helium 151

Main sequence stars 150
moments of inertia, 5, 33, 34, 88, 92, 104, 137

Neutron induced fission 18
    stars 150
    single particle energies 52
Nilsson model 48
    level scheme 50, 51
nuclear binding energy 3, 8, 121
    half lives 9
    mean field 41
    radius 22
    rotations 33, 102
    vibrations 28

Occupation number representation 99

Pairing correlations 92
    energy 26, 93, 98, 108, 121
    gap 93, 95, 98
    vibrations 101
periodic table of elements 148
potential energy contours 128

Quadrupole–quadrupole force 90
quasiparticles 99

Radius, nuclear 22
rotations 33, 102
rotation bands 11, 37, 111, 137

Seniority 93
shell effects 27, 114, 143

shape isomer 110, 120
short range correlations 43
single particle energies 41, 46, 53, 117, 131, 141
spherical shell-model 45
stability, nuclear 1, 112
 of superheavy elements 147
superheavy nuclei 141
surface energy 25, 116, 124, 154
symmetry energy 26

Tamm-Dancoff approximation 84
two centred shell-model 40, 127, 130

Vibrations, nuclear 28

White dwarf star 150
W.K.B. approximation 138